Physical Science for Engineers

Second Edition

Physical Science for Engineers

Second Edition

Walter Schofield
Vice Principal, Hackney College

McGRAW-HILL Book Company (UK) Limited

London . New York . St Louis . San Francisco . Auckland . Bogotá . Guatemala . Hamburg
Johannesburg . Lisbon . Madrid . Mexico . Montreal . New Delhi . Panama . Paris . San Juan
São Paulo . Singapore . Sydney . Tokyo . Toronto

Published by

McGRAW-HILL Book Company (UK) Limited
MAIDENHEAD . BERKSHIRE . ENGLAND

British Library Cataloguing in Publication Data
Schofield, Walter
　Physical science for engineers.—
　2nd ed
　1. Physics
　I. Title
　530'.0246　　QC21.2　80-42109
ISBN 0-07-084642-1

Library of Congress Cataloging in Publication Data
Schofield, Walter
　Physical science for engineers.
　Includes index.
　1. Physics. 2. Chemistry, Physical and
　theoretical. 3. Engineering. I. Title.
　QC21.2.S35　1981　　530　　81-8176
ISBN 0-07-084642-1　　　　　　AACR2

12345　　83210

Filmset in 'Monophoto' Times New Roman by Eta Services (Typesetters) Ltd., Beccles, Suffolk

PRINTED AND BOUND IN GREAT BRITAIN

Contents

About this Book

This book provides a background of physical science for engineering and science technicians. It is written to meet the teaching objectives of the Technical Education Council and to form a basis for further study in both science and technology.

All too often the shortage of time limits course-work to a minimum. A textbook is not so restricted and can help by providing more explanation and applications of the basic concepts. The wider coverage allows lecturers and students to select material relevant to a particular industrial background.

At the same time students must know exactly what is required of them. To this end the learning objectives for standard TEC units appear in the appendix with page references. Self assessment tests including more than 400 multiple choice questions and numerical exercises feature strongly throughout.

Diagrams and illustrations describe engineering and science much more vividly than words. I have used illustration wherever possible with the aim of keeping the text lively and readable.

My previous text *Physics for Engineers* found favour with many readers who were not taking technical courses. Although a little knowledge of mathematics is helpful, it is not essential, and non technical readers may find this book a useful introduction to applied science.

It may be that technical courses need the reinforcement of general studies to provide a balanced education. Too often perhaps we do not reveal that technology is rich in human values and controversial issues, which are highly relevant to our everyday lives. When it is well taught, technology provides a vehicle for a balanced education which is second to none. I hope that this textbook will encourage an appreciation of both the liberal and technical aspects of physical science.

TEC Units covered
by this book

Physical Science I   TEC U 75/004 and U 80/682
Engineering Science I   TEC U 76/064
Physics I   TEC U 76/002
Related Science II   TEC U 75/015
Physical Principles A   TEC 77/CUS/02

Acknowledgements

Dave Gilborn assisted with the problems
Carl Schofield assisted with the diagrams
Mildred produced the typescript

Reproduction Drawings Ltd, of Sutton,
prepared the artwork

1.
Basics

Being scientific is not just knowing a collection of facts, it is a state of mind. Therefore do not postpone being scientific until the distant day when you are qualified. Approach science with an enquiring and demanding attitude and the evidence and theory will become part of your common sense. Accept what a lecturer or a textbook says without question and although you may learn, you are unlikely to understand fully and after your examination the benefit will not stay with you long.

Modern man has achieved much by his discovery and application of science. In so doing he stands on the shoulders of individual geniuses and organized societies that came before. The Greeks and Arabs laid the foundations of subjects we study today. They made great strides in logic and mathematics and they constructed buildings which have never been excelled in beauty. The Romans were fine engineers and constructed superb roads, bridges, water supplies and systems of sanitation. (Figs. 1.1 and 1.2.)

Not so very far from Rome in 1500 A.D. lived a man who combined the abilities of scientist, engineer, and artist in a way never equalled before or since. His name was Leonardo da Vinci. He described a possible diving helmet, portable bridges, a flying machine and a parachute. (Fig. 1.3.) He hated war, although he invented several guns and an armoured car, and he would not publish the details of his submarine for fear it might be used to sink ships. He invented roller bearings, a sprocket chain, a screw-cutting lathe, and many other things, but despite all of these he is remembered most for his artistic work, notably his painting of the Mona Lisa.

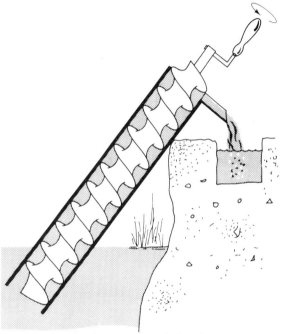

Fig. 1.1 Archimedes screw was designed to pump water for irrigation

The Scientific Method
The Rules of the Game

The fantastic progress made by science in the last century or so, has been brought about by the application of a sequence of activities known as the scientific method:

1. Observation
2. Experiment
3. Analysis
4. Prediction
5. Verification

Fig. 1.2   A model of the architecture of ancient Rome
(*Courtesy of The Italian State Tourist Office*)

It is a cyclic process which goes on and on establishing facts and relationships.

1. Observation   Any event or experiment must be subjected to careful and systematic scrutiny, and not simply given a passing glance.

Fig. 1.3   Da Vinci's design for a helicopter

2. Experiment   By this means we verify phenomena under controlled conditions, and establish that we are not dealing with an illusion or an effect of coincidence.

3. Analysis   We search for a pattern into which all observations can be fitted.

4. Prediction   We examine the possible consequences of the deductions in 3, and apply them to other circumstances.

5. Verification   All predictions must be checked by further experiment.

The Evidence

An observed relationship is only accepted as scientific fact if it is capable of verification, i.e., if the conditions are reproduced exactly then the observed effect is repeated also. If we cannot repeat it then we have not accounted for all the variables and we must keep an open mind about the evidence. Thus the existence of unidentified flying objects (flying saucers) cannot be accepted as a scientific fact until they are observed under controlled conditions. It would however, be quite unscientific to deny their existence.

When our knowledge of an individual phenomenon is incomplete, we may yet establish useful relationships based on a large number of occurrences. For example we have no way of predicting the precise moment that a particular radioactive atom will emit a particle, but the rate of emission from a kilogram of the substance can be predicted with great accuracy. In the same way, well balanced dice can be relied on to give about 1000 sixes in 6000 throws.

We also apply statistical methods to situations which are too complex to define in detail. Thus the individual movements of the molecules in a gas are too numerous to plot, but we may deal with their group properties, such as pressure and temperature with some accuracy. Weather forecasting is a case where probabilities are applied to very complex large scale occurrences with a less certain but still useful success rate.

We must be careful to avoid the development of superstition. Men have always been prone to this when they ignore the laws of chance and probability. Superstitions arise when two impressive events coincide by chance. Thus a rain dance which is followed by rain is remembered while the unsuccessful dance at another time is forgotten. The connection becomes a matter for belief, and one more superstition is born.

On the other hand many discoveries were made by chance long before they were confirmed by systematic analysis. Mould was used to dress wounds before the discovery of penicillin, which is derived from a mould. Many other folk cures have a basis in fact.

In science we are trying to make a pattern out of our experiences and our measurements. Sometimes more than one pattern will contain the facts and then we choose the simplest pattern. The simplest pattern is not just the one which is the briefest or the easiest to explain. It is the one which is the most easily challenged or confirmed

by new evidence. For example, we may say that thunder is due to the anger of the gods or due to the discharge of electrons from a cloud. The former defies contradiction (and supporting evidence) while the latter is capable of being confirmed or disproved.

Towards Objectivity

The scientist must bear in mind that, in applying the scientific methods to his investigations, he is still an animal and, like all animals, is limited by his own fallible and unreliable senses. While his brain is the most wonderful organism in creation, it depends for information on messengers that sometimes deliver garbled and inaccurate messages. Your eyes, for example, convey one impression of the pattern in Fig. 1.4 which a few

Fig. 1.4 Optical illusion associated with the sharp contrast of light and shade

measurements made with a ruler will correct. This is an example of an optical illusion, but it is not only the eyes that make mistakes. If you put your right hand in hot water at the same time as you put your left hand in cold, and then plunge both hands simultaneously into a bowl of luke-warm water, your right hand will tell you the water is cold, and your left hand that the water is hot. Some method must obviously be found of accurately informing the brain of the actual measurements of length, mass, temperature, and time which relate to any particular investigation. Scientists and Engineers are consequently trying to find new methods for measuring the effects of physical phenomena, and to make the instruments more accurate.

Scientists are always striving to make observations which are not dependent on the observer. For example, the sun appears to go around the earth but further analysis shows that the earth rotates on its axis daily and in fact goes around the sun annually. Another form of subjectivity is that in which the process of measurement interferes with the quantity being measured. An ammeter placed in a circuit will record a current but because of its resistance it will affect the current in the circuit. Another example of subjectivity would be using a mercury thermometer to measure the temperature of a thimble full of liquid because its temperature would be changed by the thermometer.

Prejudice has been a great stumbling block to the progress of science. People form a theory which satisfies them on incomplete evidence, and then stick to it in spite of new evidence. For example, the idea that the stars were embedded in several crystal domes which rotated around the earth as centre of the universe became a basic belief of the early Christian church. Galileo made observations which contradicted this but his evidence was suppressed and he was kept under permanent house arrest. Even today there are still States in America where Darwin's theory of evolution may not be taught because the authorities think it challenges the Bible.

Of course scientists must not close their minds to supernatural phenomena. Everything from extra sensory perception to witchcraft and religious hypotheses should be examined in the light of the evidence available. Faith in analysis of facts is out of place. It is only a virtue when we need courage to proceed with convictions which have arisen from our analysis of the facts.

The progress of science has been retarded during certain periods in man's history when it was held to be wrong to challenge the authority of the government, the church, and the opinions of certain ancient scientific writers. Even worse than this have been those periods when scepticism was regarded as a sin, and publishing factual discoveries a punishable crime. We hope, now that science is quite respectable, that this stage in the development of science is over. You can find a similar situation, however, in present day social and economic science which is in its infancy. The belief that public ownership leads more efficiently to the things which man values, is regarded as a crime in some Capitalist countries, just as the belief that

private ownership is more efficient is a crime in some Communist countries.

In human society where disagreement generates aggression then violence may soon follow. Democratic processes are on trial as a means of containing and resolving human conflict, but should they fail then more authoritarian systems may replace them. One day we may come to regard these questions as a matter for objective experiment.

The International System

The International system of units is based on the metre, the kilogram and the second and also on three other units; the kelvin (temperature), the ampere (electric current), and the candela (light illumination). These units are all referred to as arbitrary units since their values were given to them by a decision and not as a result of a deduction. (Table 1.1.) Other units can be defined

Table 1.1 Standard International (SI) Units

Quantity	Name of unit	Unit symbol
Length	metre	m
Mass	kilogram	kg
Time	second	s
Electric current	ampere	A
Thermodynamic temperature	Kelvin	K
Luminous intensity	candela	cd

in terms of these basic units without any further arbitrary decisions. They are called derived units to distinguish them from the basic units.

In some cases we need to deal in multiples or submultiples of units. A complete range each with its special name is given in Table 1.2, but in practice only a few of them are used. It is considered good form to use only multiples of 1000, e.g., metres, millimetres (1/1000 m), and kilometres (1000 m), and not the centimetre (1/100 m), or decimetre (1/10 m).

Measurements of any kind ultimately depend on a comparison with a basic quantity. The accuracy of any measurement we make depends on the precision with which the comparison can be made. Just as you cannot measure a distance accurately

Table 1.2   Standard Multiplication Factors

Multiplication factor	Power	Prefix	Symbol
1 000 000 000 000	10^{12}	tera	T
1 000 000 000	10^9	giga	G
1 000 000	10^6	mega	M
1 000	10^3	kilo	k
100	10^2	hecto	h
10	10	deca	da
1			
0.1	10^{-1}	deci	d
0.01	10^{-2}	centi	c
0.001	10^{-3}	milli	m
0.000 001	10^{-6}	micro	μ
0.000 000 001	10^{-9}	nano	n
0.000 000 000 001	10^{-12}	pico	p

with an inaccurate ruler, so no measurement can be made more accurately than that achieved in its standardization. Thus we go to infinite pains to preserve the constancy of our standards and to improve our methods of comparison.

Length

With their strong sense of reason the French chose the unit of length as a certain fraction of the earth's circumference. With their even stronger sense of nationalism they chose the 1 ten millionth of the distance from the equator to the North pole through their own capital city Paris. They called it the metre. Comparisons with such a distance would not be convenient to say the least and for a hundred years or more the metre was defined in terms of a platinum-iridium bar kept, like the standard kilogram, in Paris. In 1961 however we switched to a standard of length which can not warp or expand and is completely indestructable. The metre is now defined as precisely 1 650 763.73 wavelengths of the orange red light emitted by the gas krypton. (Fig. 1.5.) The multiples and submultiples of the metre are given in

1 m = 1 650 763.73
wavelengths

Fig. 1.5   The standard metre is defined in terms of the wavelengths of krypton light

Table 1.3. From the basic units of length we derive a unit of area, the square metre, and of volume, the cubic metre. This unit of volume is rather large and we sometimes use the litre ($= \frac{1}{1000}$ cubic metre) and millilitre ($= 10^{-6}$ cubic metres).

Table 1.3

Unit	Symbol	Definition
Kilometre	km	1000 m
Metre	m	
Millimetre	mm	10^{-3} m
Micrometre	μm	10^{-6} m
Nanometre	nm	10^{-9} m

Mass

The basic unit of mass is the kilogram. It is the mass of a particular piece of platinum kept in the Bureau of Weights and Measures in Paris. The standard kilogram was originally made to match the mass of one thousand cubic centimetres of pure water at $0°C$. Once made however it is the platinum and not the water which has become the standard.

Table 1.4   Mass Units

Name	Symbol	Definition
Kilogram	kg	1 kg
Gram	g	10^{-3} kg
Milligram	mg	10^{-6} kg
Microgram	μg	10^{-9} kg

The concept of mass needs careful thought. Mass is a measure of the inertia or reluctance of a free body to move when a force is applied to it. Thus when we say a body has a mass of 1 kg we mean that it accelerates to the same extent as the standard platinum kilogram when the same force acts on it.

Be careful to distinguish mass from weight which is really a force. Weight is equal to the force exerted on a body when it is supported relative to the earth. Weight is dependent on the environment, for example it is greater at the poles than at the equator because the poles are nearer to the centre of the earth than the equator. Weight is also affected by the earth's rotation and the density of the surface rock in a particular locality. On the moon the force of gravity on a man is reduced to 1/6 of its earth value as a spring balance would show. Sometimes an object appears to have no weight at all relative to an orbiting or free falling space craft. The mass of a body however remains constant in each of these situations as a beam balance would show. (Fig. 1.6.)

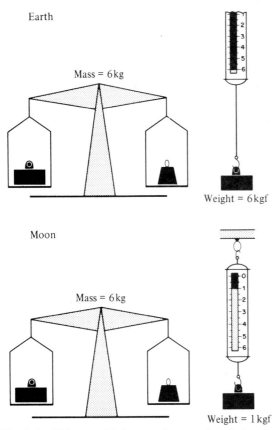

Fig. 1.6 Objects weigh less on the moon but their mass stays the same

Once you have grasped the idea of the difference between mass and weight, relax. Any kind of balance you use to weigh an object registers the mass directly in mass units, i.e., kilograms. Any variations of weight on the earth are too small to be significant.

Time

To establish a standard of time, we require a process which is repeated endlessly and with ab-solute regularity. The most obvious recurrent phenomenon in human experience is the fact that day follows night, and our sense of time stems from the alternation of night and day and of winter and summer. The day is divided into hours, minutes and finally seconds. The second, our basic unit of time was originally defined as a fraction of a day, or to be precise $\left\{\frac{1}{60} \times \frac{1}{60} \times \frac{1}{24}\right\}$ of a mean solar day. This is the average time for the earth to rotate on its axis as measured by a sundial, but unfortunately it is not constant. The viscous effect of the tides slows down the rotation and increases the year by $3\frac{1}{2}$ seconds every 1000 years!

For a long time since its discovery by Galileo the nearly constant time of swing of a pendulum has played an important part in the measurement of time. However, the swing of the pendulum is subject to factors such as variations in the temperature and the viscosity of the air, which influence slightly the length of time it takes to swing. An oscillation of a different kind now provides us with our standard. It is that of the caesium atom which changes from one arrangement of its electrons to another and back again at an amazingly constant rate. The present definition of the second is the duration of 9 192 631 770 periods of the radiation from the transitions between the two levels of the ground state of the caesium atom. Clocks based on this type of atomic vibration are accurate to within a second in 300 years. Quartz clocks and watches use a similar oscillation of a tiny crystal of quartz (calcium carbonate). Cheap and compact quartz watches produce an accuracy of about 10 seconds a month. Just a few years ago an expensive and bulky chronometer would be needed to achieve this accuracy.

Density

Density is an important basic property of materials, for example, it is one of the things that makes gold different from polystyrene.

$$\text{Density} = \frac{\text{mass}}{\text{volume}}$$

The symbol used for density is the Greek letter ρ

$$\rho = \frac{M}{V}$$

The units should always be expressed in kg/m^3. Just to give you some idea of the order of size of densities, water has a density of $1000 \, kg/m^3$. In other words a cubic metre of it has a mass of 1000 kg (a metric ton). Copper has a much higher density and a cubic metre of copper has a mass of 8900 kg.

Now the density of water is a rather important reference point for man, after all $\frac{2}{3}$ of the world is covered by it! If you fall in the deep end of a swimming pool, it matters that your density is just less than water and therefore you float. Almost as important is the fact that when you drop your keys overboard their density is greater than that of water and they sink without trace. Because we have this inbuilt notion of how dense water is we find it useful to express other densities in terms of the density of water.

$$\text{Relative density} = \frac{\text{density of material}}{\text{density of water}}$$

Another way of expressing this is

$$\text{Relative density} = \frac{\text{mass of substance}}{\text{mass of same volume of water}}$$

Since relative density is a ratio of two quantities with the same dimensions it has no units. Note that thermal expansion causes a small reduction in densities and therefore values are quoted for a stated temperature. (Table 1.5.)

Table 1.5 Densities and Relative Densities

	Density (kg/m^3) at 20°C	At 20% relative density
Water (4°C)	1×10^3	1.0
Aluminium	2.7×10^3	2.7
Copper	8.9×10^3	8.9
Gold	19.3×10^3	19.3
Iron	7.7×10^3	7.7
Lead	11.3×10^3	11.3
Brass	8.5×10^3	8.5
Mercury	13.5×10^3	13.5
Oil	0.9×10^3	0.9
Sea water	1.025×10^3	1.025
Air at 0°C and standard atmospheric pressure	1.29	0.00129
Expanded polystyrene, approximately	20	0.02

Problem

A water tank in the form of a cylinder 0.8 m diameter is filled to a depth of 0.6 m. Calculate the mass of water. (Fig. 1.7.)

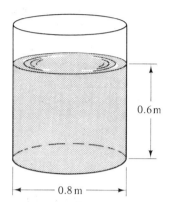

Fig. 1.7 A cylindrical water tank

Solution

$$\text{Density} = \frac{\text{mass}}{\text{volume}}$$

$$\text{Mass} = \text{volume} \times \text{density}$$
$$= \pi \times (0.4)^2 \times 0.6 \times 1 \times 10^3$$
$$\text{Mass of water} = 301.4 \, kg$$

Problem

An iron pipe 5 m long has an internal diameter of 250 mm and a wall thickness of 6 mm. (Fig. 1.8.) Calculate its mass if the iron has a density of $7.8 \times 10^3 \, kg/m^3$.

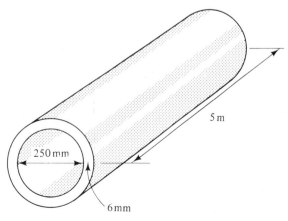

Fig. 1.8 Cylindrical pipe

Solution

Mean diameter of pipe
$$= 250 + 6 = 256 \text{ mm} = 0.256 \text{ m}$$
Area of cross-section of iron
$$= \pi \times \text{mean diameter} \times \text{thickness}$$
$$= 3.14 \times 0.256 \times 0.006$$
$$= 48.2 + 10^{-3} \text{ m}^2$$
Volume = length × cross-sectional area
$$= 5 \times 4.82 \times 10^{-3}$$
$$= 24.1 \times 10^{-3} \text{ m}^3$$
Mass = volume × density
$$= 24 \times 10^{-3} \times 7.8 \times 10^3$$
Mass of pipe = 188 kg

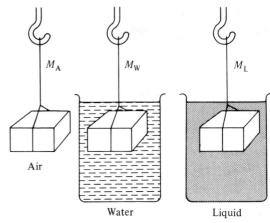

Fig. 1.9   Measurement of the density of solids and liquids

Measurement of Density

The volume of an object of regular shape like a cylinder or rectangular solid can be calculated from its measurements. The density is then calculated from its mass obtained by weighing

$$\text{Density} = \frac{\text{mass}}{\text{volume}}$$

Irregular shapes require craftier methods. When an object is immersed in a liquid its weight is reduced because of the bouyant effect of the liquid. The reduction in weight is equal to the weight of liquid displaced (the famous principle of Archimedes). We use this as the basis of a very simple and convenient method of measuring density.

An object is weighed in air M_A and then weighed again suspended in water M_W. (Fig. 1.9.)

$$\text{Relative density} = \frac{\text{mass of an object}}{\text{mass of same volume of water}}$$

$$= \frac{M_A}{M_A - M_W}$$

To calculate the actual density from the relative density multiply by the density of water, i.e., 10^3 kg/m^3.

Problem

An object weighs 24.5 g in air and 20.2 g in water. Calculate its relative density and density.

Solution

$$\text{Relative density} = \frac{M_A}{M_A - M_W} = \frac{24.5}{24.5 - 20.2} = 5.69$$

$$\text{Density} = 5.69 \times 10^3 \text{ kg/m}^3$$

Relative density of object is 5.69 and its density is $5.69 \times 10^3 \text{ kg/m}^3$.

The same principle can be used to measure the density of liquids by weighing an object in air M_A in water M_W and in the liquid M_L

Relative density of liquid

$$= \frac{\text{mass of liquid}}{\text{mass of same volume of water}}$$

The volume is that of the object although we do not need to know its actual value.

Relative density of liquid

$$= \frac{\text{weight loss in liquid}}{\text{weight loss in water}}$$

$$= \frac{M_A - M_L}{M_A - M_W}$$

Problem

An object weighs 50 g in air, 43.1 g in water and 42.4 g in a liquid. Calculate the relative density and density of the liquid.

Solution

$$\text{Relative density of liquid} = \frac{M_A - M_L}{M_A - M_W}$$

$$= \frac{50 - 42.4}{50 - 43.1}$$

$$= \frac{7.6}{6.9} = 1.1$$

$$\text{Density} = 1.1 \times 10^3 \text{ kg/m}^3$$

The liquid has a relative density of 1.1 and a density of 1.1×10^3 kg.

Another way of measuring density is to use a density bottle which has a fine aperture allowing it to be filled accurately each time with the same volume of liquid. (Fig. 1.10.)

$$\text{Relative density} = \frac{\text{mass of liquid}}{\text{mass of same volume of water}}$$

$$= \frac{M_L - M_E}{M_W - M_E}$$

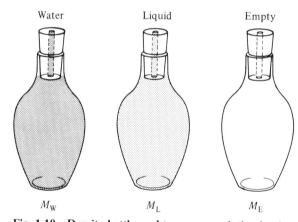

Water   Liquid   Empty

M_W M_L M_E

Fig. 1.10   Density bottle used to measure relative density

Hydrometers

Quality control of a product or of a process may call for the measurement of liquid density over a limited range but to a high degree of accuracy. For example, milk or beers or wines may have to be within certain density limits to comply with the law. In these cases a hydrometer as shown in Fig. 1.11(a) gives both convenient and accurate measure of density.

It is simply a weighted glass tube which floats in the liquid under test. The denser the liquid, the higher the hydrometer floats. By having a narrow stem a large movement occurs for small changes in density and the scale can be calibrated accurately.

The density of the acid determines the state of discharge of a lead acid accumulator. Liquid is drawn from the battery as shown in Fig. 1.11(b).

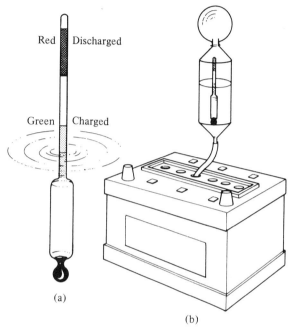

Red   Discharged

Green   Charged

(a)

(b)

Fig. 1.11   Hydrometer used to measure the density of liquids

Coherent Units

Man has developed meaningful ways of describing size in such expressive language as a days' march, or a midges' whisker. These particular distances are near the extremes of our ability to visualize. Beyond them distances become large numbers or small fractions difficult to imagine though just as easy to deal with mathematically. Outside the range of our common sense we can go with confidence providing we adopt a consistent procedure:

1. Design equations and formulae which are valid for a set of basic units.
2. Express all quantities in these units before insertion in the formulae.

Thus, although we may talk about millimetres or kilometres generally, the only units used in basic calculation are metres. Car speedometers calibrated in km/h may make sense but the only unit of velocity to use in basic formulae is the metre/second. The set of units built up from the basic units of the metre, the kilogram and the second, etc., are called coherent units. (Table 1.6.) Know which units are coherent and express all quantities in them before insertion in basic formulae.

Table 1.6 Derived SI Units

Physical quantity	SI units	Unit symbol
Area	square metre	m^2
Volume	cubic metre	m^3
Density	kilogram per cubic metre	kg/m^3
Velocity	metre per second	m/s
Acceleration	metre per second squared	m/s^2
Pressure	pascal (N/m^2)	Pa
Thermal conductivity	watt per metre Kelvin	$W/(m\ K)$
Force	newton	N $= kg\ m/s^2$
Work, energy quantity of heat	joule	$J = N\ m$
Power	watt	$W = J/s$
Electric charge	coulomb	$C = A\ s$
Frequency	hertz	$Hz = s^{-1}$
Luminous flux	lumen	$lm = cd\ sr$
Illumination	lux	$lx = lm/m^2$

Non-coherent units which are multiples or fractions of the coherent units may be used in conversation or in statements of results but even then stick to those which are related to the coherent units by factors of 1000. (Table 1.2.) As for the pounds, the miles and the inches, they will survive a generation or two and then join the guineas, the fathoms and the leagues, and live on only in poems and pantomimes.

Questions

1. Why can no measurement ever be more accurate than the accuracy with which the basic unit is defined?

2. What is the difference between a basic and a derived unit?

3. Why are scientific questions settled by experiment but political questions are kept open by argument?

4. How can we prevent an impressive coincidence giving rise to superstitious belief?

5. What has the scientific method got to offer you when you are frightened in the dark?

Multiple Choice Questions

1. Which prefix means a millionth part?
(a) mega, (b) kilo, (c) milli, (d) micro.

2. Which unit is defined in terms of light waves?
(a) mass, (b) current, (c) length, (d) time.

3. Which is not an international basic unit?
(a) litre, (b) second, (c) metre, (d) kilogram.

4. What are the units of density?
(a) kg/m, (b) kg/m^2, (c) m/kg, (d) kg/m^3.

5. What is the formula for the density of a cylinder of mass m, height h and radius r?

(a) $\dfrac{mr}{\pi m}$, (b) $\dfrac{m}{\pi r^2 h}$, (c) $\dfrac{m}{\pi rh}$, (d) $\dfrac{\pi r^2 h}{m}$.

6. Relative density of iron means its density relative to
(a) air, (b) steel, (c) platinum, (d) water.

7. Three kilograms of gold and three kilograms of expanded polystyrene have the same
(a) density, (b) volume, (c) mass,
(d) force exerted by gravity.

8. Which of the following would 1 kg of water at 0°C have greater than 1 kg of ice at 0°C?
(a) mass, (b) density, (c) volume, (d) relative density.

9. A hydrometer is a device to measure
(a) pressure, (b) force, (c) weight, (d) density.

10. The thickness of each page in this book is approximately
(a) 1.3×10^{-1} m, (b) 1.3×10^{-2} m, (c) 1.3×10^{-3} m, (d) 1.3×10^{-4} m.

Problems

1. Calculate the volume and surface area of a cube of side 1.20 m. [1.73 m^3; 8.64 m^2]

2. Calculate the volume of a cylinder of length 1.5 m and radius 0.020 m. [$1.8(8) \times 10^{-3}\ m^3$]

3. If the cube in question 1 is made of polystyrene, what would be its mass? (Density of polystyrene is 20 kg/m^3.) [34.6 kg]

4. Calculate the mass of air in a room measuring 4.0 m $\times$ 5.0 m $\times$ 2.5 m. (Density of air is 1.3 kg/m^3.) [65 kg]

5. A cylindrical water tank has a base of area 0.6 m^3. What mass of water does it hold when the water is 0.8 m deep? (Density of water is 1000 kg/m^3.) [480 kg]

6. A rod of square section measures $2.0 \text{ m} \times 0.030 \text{ m} \times 0.030 \text{ m}$. Its mass is 14 kg. Calculate its density and relative density. (Density of water is 1000 kg/m^3.) [7780 kg/m^3; 7.78]

7. A tank is made of 6 mm thick steel plate. It measures $2.0 \text{ m} \times 0.8 \text{ m} \times 1.0 \text{ m}$ high and has no lid. Calculate its mass if the density of steel is 7.8 Mg/m^3. What would be its total mass if it were filled with oil of relative density 0.80? (Density of water is 1000 kg/m^3.) [340 kg, 1620 kg]

8. A rectangular block has a mass of 6300 kg and is made of material of density 2800 kg/m^3. The area of the base is 9.0 m^2. Calculate the height of the block. [250 mm]

Multiple Choice Answers

1(d), **2**(c), **3**(a), **4**(d), **5**(b), **6**(d), **7**(cd), **8**(bd), **9**(d), **10**(d).

2.
Dynamics

Movement may be fantastically complex as of a running animal or simple as of a spacecraft in flight. Animals and humans have a highly developed concept of movement which is evident in the hawk catching its prey or in a circus juggler at work. (Fig. 2.1.) Outside the range of our instinctive co-ordination we have to break down motion into its simplest form in order to analyze it precisely.

Fig. 2.1 Man and animals possess a highly developed sense of motion

Distance and Displacement

A walk from home to a point a kilometre to the north and back represents a distance travelled of 2 kilometres, but a final displacement of zero! Distance describes the length of the journey. Displacement specifies the final change in position in both magnitude and direction. Thus after $1\frac{1}{2}$ kilometres of the journey, the distance travelled is $1\frac{1}{2}$ kilometres and your displacement is $\frac{1}{2}$ kilometre due north.

Problem

A man runs half way round a circular track of diameter 200 m starting at the most easterly point. What distance has to be covered and what is his displacement? (Fig. 2.2.)

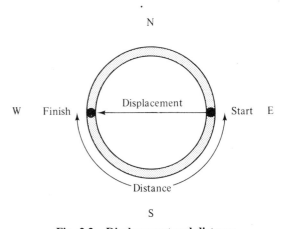

Fig. 2.2 Displacement and distance

Solution

$$\text{Distance} = \frac{\pi \times \text{diameter}}{2} = \frac{3.14 \times 200}{2}$$

$$= 314 \text{ m}$$

$$\text{Displacement} = 200 \text{ m due west}$$

The distance covered is 314 m and the displacement is 200 m due west.

Speed

Almost as important as what happens is how quickly it happens, i.e., the rate of change with respect to time or more briefly, the rate of change.

You probably know what speed looks like and what it feels like, but can you define it?

Speed is the rate at which distance is covered, i.e.,

$$\text{Speed} = \frac{\text{distance moved}}{\text{time taken}} \quad \frac{\text{(metres)}}{\text{(seconds)}}$$

Resist the temptation to use ye old British units. You may drive to work on miles per hour but once there think of speed in metres per second (m/s): one metre per second is a slow stroll, 3 m/s a comfortable jog, 14 m/s the speed limit in town, and about 30 m/s is the limit on the motorway.

During a journey, even on a motorway, the speed of a car varies. What we calculate from the above expression is the average speed during the distance and time considered. To calculate an instantaneous speed the time interval used must be short. In the case of constant speed then the average speed and the instantaneous speed are the same.

Problem

A train completes the run from Manchester to London, a distance of 270 km in 2 hours 35 minutes. Calculate its average speed in km/h and in m/s.

Solution

$$\text{Average speed} = \frac{\text{distance travelled}}{\text{time taken}}$$

$$\text{Average speed (m/s)} = \frac{270 \times 1000}{(2 \times 60 + 35)60}$$

$$= \frac{270\,000}{155 \times 60} = 29.0$$

$$\text{Average speed (km/h)} = \frac{270}{155/60}$$

$$= 104.5 \text{ km/h}$$

Average speed of train is 29 m/s
or 104.5 km/h

Speed and Velocity

In everyday language we use the words speed and velocity to mean the same thing. The units are the same so what then is the distinction between them?

Speed is the rate of change of distance, i.e.,

$$\text{Speed} = \frac{\text{distance travelled}}{\text{time taken}}$$

Velocity is the rate of change of displacement, i.e.,

$$\text{Velocity} = \frac{\text{displacement}}{\text{time}}$$

$$= \frac{\text{distance travelled in a defined direction}}{\text{time taken}}$$

With velocity the direction is specified whereas with speed it is not. So we can think of velocity as the speed in a defined direction. What happens in practice is that in a situation where direction does not enter into a calculation we use either speed or velocity, but where the direction counts we use only velocity.

The road distance to Brighton from London is 80 km but the displacement of Brighton from London is 74 km due south. For a vehicle which covers the distance in 90 minutes:

$$\text{Average road speed} = \frac{80}{90 \times 60} \times 1000 = 14.8 \text{ m/s}$$

$$\text{Average velocity} = \frac{74 \times 1000}{90 \times 60}$$

$$= 13.7 \text{ m/s due south}$$

Acceleration

When a sports car zooms ahead and leaves other cars standing at the traffic lights, we say it has a high acceleration. It has increased its velocity in a short interval of time.

Acceleration is defined as the rate of change of velocity.

$$\text{Acceleration} = \frac{\text{change in velocity}}{\text{time taken}}$$

Its units are metres per second per second, i.e., m/s^2.

Problem

A car accelerates uniformly from a standing start to 90 km/h in 8 seconds. Calculate its acceleration in metres per second per second.

Solution

$$90 \text{ km/h} = \frac{90\,000}{60 \times 60} = 25 \text{ m/s}$$

$$\text{Acceleration} = \frac{25}{8} = 3.125 \text{ m/s}^2$$

When an object reduces its velocity we say it decelerates, or if we are dealing with it mathematically we may say it has a negative acceleration.

If an object starts with a velocity u and accelerates uniformly to a velocity v in t seconds we can calculate its acceleration a as:

$$a = \frac{v - u}{t}$$

or re-arrange this to give the final velocity v

$$v = u + at$$

Problem

An aeroplane taxying at 8 m/s accelerates at 5 m/s^2 for 10 seconds, what velocity does it reach?

Solution

$$v = u + at$$
$$v = 8 + 5 \times 10$$
$$\text{Velocity reached} = 58 \text{ m/s}$$

If we wish to calculate the distance travelled s by an object during a uniform acceleration from velocity u to velocity v, we can take the average velocity to be half way between u and v, i.e., $(u+v)/2$. For example, during a uniform acceleration from 4 m/s to 8 m/s an object's average velocity is 6 m/s. Then distance = average velocity × time

$$s = \frac{(u+v)}{2} \times t$$

Problem

Calculate the distance travelled by the aeroplane in the above problem during the acceleration.

Solution

$$s = \frac{(u+v)}{2} t$$

$$= \left(\frac{8+58}{2}\right) \times 10 = 330 \text{ m}$$

Distance travelled by aeroplane is 330 m.

Distance–Time Graphs

Car speedometers are notoriously inaccurate instruments and even the law allows them to have a 10 per cent error. We can measure speed more accurately by timed runs over a measured distance. In the case shown in Fig. 2.3 the speed was steady at 20 m/s and the time was recorded at 100 metre intervals.

Plotting the motion on a distance–time graph gives a picture of the run as it progresses. (Fig. 2.4.) The gradient of the line is obtained by dividing the height of the triangles in metres by the base in seconds.

$$\text{Gradient} = \frac{100 \text{ metres}}{5 \text{ seconds}} = 20 \text{ m/s}$$

Notice that the gradient is equal to the speed. Because the speed is constant the gradient is constant and the graph is a straight-line. Figure 2.5

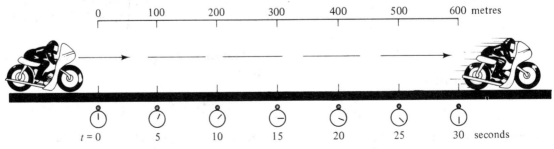

Fig. 2.3   Timed run at constant speed

14

shows how the gradient differs for higher and lower uniform speeds.

When a car accelerates from a standing start the distance time graph is a curve. As the speed increases so the gradient increases. The top speed is reached when the gradient is a maximum and the line is straight. (Fig. 2.6.)

To measure the speed at any instant we measure the gradient of the graph at that point as follows:

1. Draw a tangent to the curve at the point

2. Construct a right-angled triangle using the tangent as the hypoteneuse
3. Measure the height of the triangle in metres
4. Measure the base of the triangle in seconds
5. Divide the metres by the seconds to give speed

If a car is clocked as it slows down as in a braking test the graph curves the other way. As the car stops the graphs becomes horizontal (zero gradient). (Fig. 2.7.)

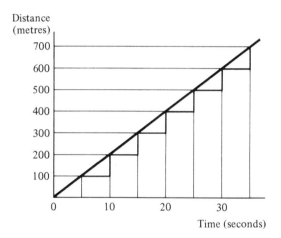

Fig. 2.4 **Distance–time graph for the constant speed run of Fig. 2.3**

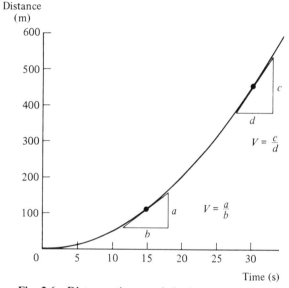

Fig. 2.6 **Distance–time graph for increasing speed**

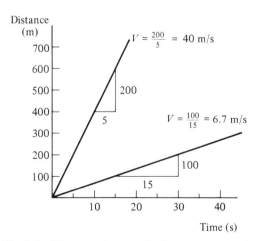

Fig. 2.5 **Distance–time graphs for constant speeds**

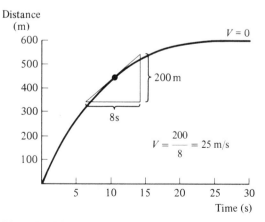

Fig. 2.7 **Distance–time graph for decreasing speed**

Velocity–Time Graphs

A constant force applied to an object produces a constant acceleraton, i.e., equal increases in velocity for equal intervals of time.

$$
\begin{array}{llllllll}
v \text{ (m/s)} & 0 & 10 & 20 & 30 & 40 & 50 & 60 \\
t \text{ (s)} & 0 & 5 & 10 & 15 & 20 & 25 & 30
\end{array}
$$

$$
\text{Acceleration} = \frac{10}{5} = 2 \text{ m/s}^2
$$

Plotted as a graph of velocity against time this gives a straight line. (Fig. 2.8.)

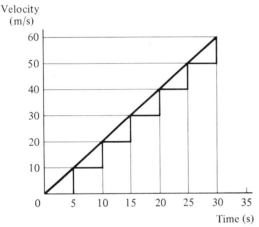

Fig. 2.8   Velocity–time graph for constant acceleration

The gradient of the graph is constant and is obtained from one of the triangles,

$$
\text{Gradient} = \frac{\text{height of triangle (in metres/second)}}{\text{base of triangle (in seconds)}}
$$

$$
= \frac{10}{5} = 2 \text{ m/s}^2
$$

i.e., the gradient of the graph is equal to the acceleration.

Figure 2.9 is a velocity–time graph of a vehicle which from a standing start

1. accelerates to 15 m/s in 10 seconds
2. maintains this velocity for 20 seconds
3. brakes to a stop in 10 seconds

We can calculate the accelerations for each section of the motion from the expression

$$
\text{Acceleration} = \frac{\text{change in velocity}}{\text{time}}
$$

$$
= \frac{v}{t} = \frac{15}{10} = 1.5 \text{ m/s}^2
$$

We could also get the acceleration from the graph, for this is the value of the gradient of the line AB in Fig. 2.9.

During the constant velocity sections represented by BC the acceleration is zero and the gradient of the line BC is also zero.

For the deceleration the change in velocity is -15 m/s.

$$
\text{Therefore } a = \frac{-15}{12} = -1.25 \text{ m/s}^2
$$

Again this is equal to the gradient of the line CD. It is negative because the line slopes downwards and represents a deceleration or negative acceleration.

For the very simple case we have chosen in Fig. 2.9 the accelerations are constant and the straight line graphs have few advantages over equations. However, no vehicle I have ever travelled in had a constant acceleration or constant deceleration. Motion is usually more complex and in these cases the graphical treatment has great advantages. Take the case of a car moving off from rest using only its bottom gear. (Fig. 2.10.)

From the graph we can see where the acceleration is greatest or where braking is most effective. We could find these maximum values of acceler-

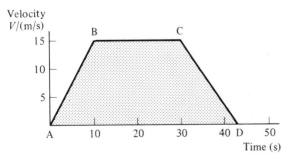

Fig. 2.9   Velocity–time graph

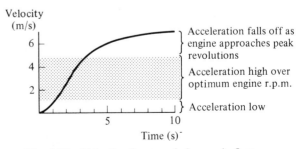

Fig. 2.10   Velocity–time graph for car in first gear

ation or deceleration by measuring the gradient at these steepest points. To find the gradient at any point we draw a tangent to the curve at that point and construct a triangle as in Fig. 2.5. We then calculate the gradient by dividing the triangle's vertical height a (in metres per second), by the base length b (in seconds).

An acceleration and braking test run, for a sports car, is represented in Fig. 2.11. Such a graph could be used to find the best points for gear changes or to detect acceleration flat spots or brake fade. What would you expect the v–t graph for a dragster to be?

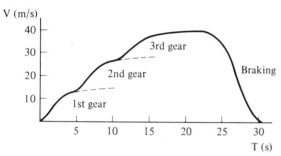

Fig. 2.11 Velocity–time graph for automobile with three gears

Problem

Sketch the velocity–time curve for an arrow which is released by a horseman at the gallop and brings down a running deer. (Solution in Fig. 2.12.)

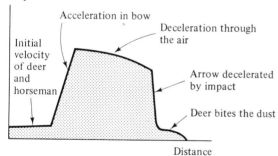

Fig. 2.12 Velocity–time graph for an arrow fired from horseback at a running deer

Free Fall

When an object is unsupported in a gravitational field, for example that of the earth, it experiences an acceleration. The value of the acceleration at the earth's surface is about 10 m/s^2. In fact it varies from place to place about the average figure of 9.81 m/s^2. It is less at the equator than at the poles and it is greater where the surface rocks are very dense.

In a vacuum the acceleration due to gravity g is independent of the mass or of the shape of an object. However, in air the resistance of the air plays a bigger and bigger part as the velocity increases. At about 180 km/h the air resistance on a falling man equals the force of gravity and his velocity remains constant. Sky divers control their speed by the position of their arms and legs. (Fig. 2.13.) A few men have survived a free fall collision with the ground. Mice jumping from an aeroplane would have a better chance of survival while elephants fortunately do not indulge in this activity.

Fig. 2.13 Sky divers free falling at about 180 km/h

Problem

An object is released from rest and falls freely under gravity. Calculate its velocity and position after 1, 5 and 10 seconds ($g = 10 \text{ m/s}^2$).

Solution

Velocity = acceleration × time

$$v = 10t$$

Thus the velocity after 1 second = 10 m/s

3 seconds = 30 m/s

6 seconds = 60 m/s

Distance = average velocity × time

$$s = \frac{(v+u)}{2}t$$

after 1 second $s = \dfrac{(10+0)}{2}\,1 = 5$ m

after 3 seconds $s = \dfrac{(30+0)}{2}\,3 = 45$ m

after 6 seconds $s = \dfrac{(60+0)}{2}\,6 = 180$ m

After 1, 3 and 6 seconds the object has fallen 5, 45 and 180 metres respectively.

Projectiles

Objects projected vertically upwards are decelerated by gravity until they stop and then they fall freely. Neglecting air resistance, the velocity on the way up at any point is equal in magnitude to the velocity at that point on the way down.

Equations of Motion

Earlier in this chapter we arrived at the equation for the distances covered by a body starting with velocity u and proceeding with a uniform acceleration a to a velocity v, during a time t.

Distance = average velocity × time

$$s = \left(\frac{v+u}{2}\right)t \tag{1}$$

also we know that final velocity = initial velocity + acceleration × time

$$v = u + at \tag{2}$$

Where any three of these quantities are known, the other two can be found by solving the equations simultaneously.

Alternatively we can derive two other useful equations.

Substitute for v from Eq. (2) into Eq. (1)

$$s = \left(\frac{u + at + u}{2}\right)t$$

$$s = \frac{2ut + at^2}{2}$$

$$s = ut + \tfrac{1}{2}at^2 \tag{3}$$

Also from Eq. (1)

$$t = \frac{2s}{v+u}$$

Substituting for t into Eq. (2)

$$v = u + a\left(\frac{2s}{v+u}\right)$$

$$v - u = a\left(\frac{2s}{v+u}\right)$$

$$(v-u)(v+u) = 2as$$

$$v^2 - u^2 = 2as$$

$$v^2 = u^2 + 2as \tag{4}$$

It is quite sufficient to use Eqs. (1) and (2) but if you also remember Eqs. (3) and (4) they will save you time when solving problems.

Problem

What initial velocity is required to project an object to a height of 80 m (take $g = 10$ m/s²)?

Solution

$$v^2 = u^2 + 2as$$

at the highest point the final velocity $v = 0$

also $a = -10$ m/s²

$$0 = u^2 + 2 \times (-10) \times 80$$
$$u^2 = 1600$$
$$u = 40 \text{ m/s}$$

The initial velocity required = 40 m/s.

Problem

How far will an object fall from rest under gravity in 4 seconds? ($g = 10$ m/s².)

Solution

$$s = ut + \tfrac{1}{2}at^2$$
$$= 0 \times 4 + \tfrac{1}{2} \times 10 \times 4^2$$
$$s = 0 + 80 = 80 \text{ m}$$
$$\text{Distance dropped} = 80 \text{ m}$$

Angular Motion

We have dealt with simple motion in a straight line but that is the exception rather than the norm. Objects under the influence of forces may rotate without moving along or more often rotate and move along at the same time.

A rotating object is said to have an angular velocity. We could measure this angular velocity in terms of complete revolutions per second or in degrees per second where 1 revolution = 360°. However it is sometimes more convenient to use a different unit called the radian.

The Radian

A radian is constructed by measuring off the radius around the perimeter of a circle. **One radian is the angle subtended at the centre of a circle by a length of arc equal to the radius.** (Fig. 2.14.)

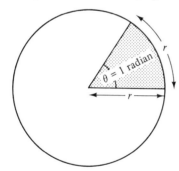

Fig. 2.14   The definition of the radian

The value of a particular angle θ in radians is obtained by dividing the length of the arc l by the radius r. (Fig. 2.15.)

$$\theta = \frac{l}{r} \text{ radians}$$

For a complete revolution the arc length is equal to the circumference of the circle, i.e., $2\pi r$

$$1 \text{ revolution} = \frac{2\pi r}{r} = 2\pi \text{ radians}$$

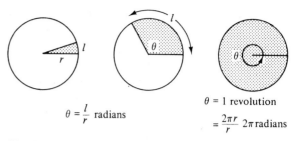

$$\theta = \frac{l}{r} \text{ radians}$$

$$\theta = 1 \text{ revolution}$$
$$= \frac{2\pi r}{r} \; 2\pi \text{ radians}$$

Fig. 2.15   To obtain the angle in radians divide the arc length by the radius

The value of a radian in degrees can be calculated from this

$$1 \text{ revolution} = 2\pi \text{ rad} = 360°$$

$$\therefore \quad 1 \text{ rad} = \frac{360}{2\pi} = \frac{360}{2 \times 3.14} = 57.3°$$

Problem

A point on the perimeter of a lathe chuck of radius 80 mm is moved 20 mm. What is the angular movement in radians?

Solution

$$\theta = \frac{l}{r} = \frac{20}{80} = 0.25 \text{ rad}$$

$$\text{Angular movement} = 0.25 \text{ rad}$$

Problem

When the lathe chuck above is rotated through 2.4 rad, how far does a point in the perimeter move?

Solution

$$l = \theta \times r = 2.4 \times 80 = 192 \text{ mm}$$
$$\text{Distance moved} = 192 \text{ mm}$$

Angular Velocity

Angular velocity is the rate of change of angle. It is measured in radians per second.

$$\text{Angular velocity} = \frac{\text{angle of rotation in radians}}{\text{time taken in seconds}}$$

$$\omega = \frac{\theta}{t} \text{ rad/s}$$

19

It is common for angular velocity to be quoted in revolutions per minute (rpm) because it is easier to visualize. For example the peak power of a motor cycle engine may be quoted for an angular velocity of 5000 rpm. In many basic equations angles are expressed in radians and so we must convert from rpm to radians, e.g.,

$$5000 \text{ rpm} = \frac{5000}{60} \text{ rps} = \frac{5000 \times 2\pi}{60} \text{ rad/s}$$
$$= 523 \text{ rad/s}$$

Angular Acceleration α

An increasing angular velocity is an angular acceleration. **Angular acceleration is defined as the rate of change of angular velocity.** It is measured in radians per second, per second. (rad/s^2.)

Angular acceleration = α

$$= \frac{\text{change in angular velocity}}{\text{time taken}}$$

$$\alpha = \frac{\omega_2 - \omega_1}{t} \text{ rad/s}^2$$

Problem
A motor cycle engine in neutral ticks over at 800 rpm and a 3 second blip of the throttle produces 6000 rpm. What is the angular acceleration?

Solution
Initial angular velocity

$$\omega_1 = \frac{800 \times 2\pi}{60} = 83.7 \text{ rad/s}$$

Angular velocity after 3 seconds

$$\omega_2 = \frac{6000 \times 2\pi}{60} = 628 \text{ rad/s}$$

Angular acceleration

$$\alpha = \frac{\omega_2 - \omega_1}{t} = \frac{544.3}{3} = 181 \text{ rad/s}^2$$

Graphical Representation of Angular Motion

By plotting graphs of angular velocity against time we can obtain a useful visual representation of angular motion. Figure 2.16 represents the motion of an object which:

(a) starts from rest and accelerates uniformly up to 1500 rpm (157 rad/s) in 5 seconds
(b) maintains this angular velocity for 15 seconds
(c) comes to rest in 10 seconds.

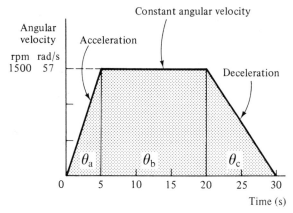

Fig. 2.16   Angular velocity–time graph

The graph is similar to that for linear motion (see page 16). The slope of the angular velocity–time graph gives the angular acceleration or deceleration. The horizontal section represents constant angular velocity and the area under the graph equals the total angle of rotation θ. For each part of the motion

$$\theta = \text{average velocity} \times \text{time}$$

$$\theta_a = \left(\frac{157}{2}\right) \text{s} = 392.5 \text{ rad}$$

$$\theta_b = 157 \times 15 = 2355 \text{ rad}$$

$$\theta_c = \frac{157}{2} \times 10 = 785 \text{ rad}$$

Total angle of rotation = $\theta_a + \theta_b + \theta_c = 3532.5$ rad

Questions
1. Suggest reasons for the fact that the force exerted by gravity reduces with both height above or depth below the surface of the earth.

2. If a man's free fall terminal velocity is 200 km/h, what would you think happened to people and smaller creatures caught in the 300 km/h winds in Hurricane Alan which swept the Carribean in August 1980?

3. You do not feel constant velocity but you do feel acceleration. Why is this?

4. Why do projectiles with the same initial speed travel highest when fired vertically? Would it be true on the moon?

5. What can you see from graphs of motion that you can't see just as well from the listed data?

Multiple Choice

1. Speed is defined as
(a) $\dfrac{\text{distance}}{\text{time}}$, (b) distance × time,

(c) acceleration × time, (d) $\dfrac{\text{acceleration}}{\text{time}}$.

2. Acceleration is
(a) the rate of change of velocity, (b) distance covered,
(c) change in velocity, (d) rate of change of distance.

3. An accelerating object has a distance–time graph which is a line
(a) parallel to the time axis,
(b) curved away from time axis,
(c) curved towards time axis,
(d) parallel to the distance axis.

4. The slope of a distance–time graph is
(a) distance, (b) velocity, (c) time, (d) acceleration.

5. The average speed of an object during a uniform acceleration from u to v in a time t is
(a) $\dfrac{v}{2}$, (b) $\dfrac{u+v}{t}$, (c) $\dfrac{uv}{t}$, (d) $\dfrac{u+v}{2}$.

6. The unit of velocity is
(a) newton, (b) metre per second,
(c) metre per second per second,
(d) newton per second.

7. The unit of acceleration is
(a) m/s², (b) N/m², (c) m/s, (d) N/m.

8. The area under a velocity–time graph is the
(a) acceleration, (b) velocity, (c) distance travelled,
(d) time taken.

9. The slope of a velocity–time graph is equal to
(a) distance travelled, (b) the velocity,
(c) the acceleration, (d) elapsed time.

10. A car is travelling south at constant speed. It passes over a piece of mud which sticks to the tyre. The initial acceleration of the mud as it leaves the ground is
(a) horizontally to the north, (b) vertically upwards,
(c) zero, (d) horizontally to the south.

Problems

1. A car travels 200 m in 25 s. Calculate its average speed. [8 m/s]

2. A skier travels a distance of 3.4 km in 20 minutes. Calculate his average speed in m/s. [2.83 m/s]

3. How long does it take for a car travelling at 95 km/hour to cover 320 km of motorway? [3 h 22 min]

4. Draw the distance–time graph for an object travelling with a constant speed of 3.0 m/s and use it to determine the distance travelled after 4.6 seconds. [13.8 m]

5. A dragster reaches a speed of 30 m/s from a standing start in 8.0 s. Calculate the acceleration. [3.75 m/s²]

6. Calculate the acceleration of a train travelling initially at 4 m/s which accelerates to 12 m/s over a half a minute. [0.27 m/s²]

7. An object falls from rest with an acceleration of 10 m/s². Plot a speed–time graph for the first 8 seconds of fall. From your graph determine the distance travelled and also the speed after 4.2 s. [320 m; 42 m/s]

8. A train uniformly accelerates from rest to 80 km/h in 2 minutes, then maintains this speed for 5 minutes before braking to rest in 15 s. Plot a speed–time graph for the journey and from it obtain (a) the acceleration, (b) the deceleration, (c) the total distance travelled. [0.185 m/s², 1.48 m/s², 8.17 km]

9. Use the equation of motion to calculate the speed and distance travelled by an object falling from rest under gravity after 3, 6 and 9 seconds. Take g = 10 m/s². [30 m/s, 45 m; 60 m/s, 180 m; 90 m/s, 405 m]

10. Confirm by calculations the results obtained graphically in problems 4, 7 and 8.

11. Calculate to the nearest metre the distance moved by an object falling from rest during (a) the first second, (b) the second second, (c) the third second. Take g = 10 m/s². [5 m, 15 m, 25 m]

12. Express the following in radians per second, (a) 500 rotations per minute, (b) 2000°/s. [52.4 rad/s, 34.9 rad/s]

13. A car has wheels of diameter 0.6 m and travels at 12 m/s. Calculate the angular velocity of the wheels. [40 rad/s]

14. A bicycle's wheels have an angular velocity of 25 rad/s. If the diameter of the wheels is 0.8 m what is the speed of the bicycle? [10 m/s]

15. If the bicycle of problem 14 reaches a speed of 8 m/s from rest in 6.2 s what is the angular acceleration of its wheels? [3.2 rad/s²]

Multiple Choice Answers

1(a), 2(a), 3(b), 4(b), 5(d), 6(b), 7(a), 8(c), 9(c), 10(b).

3.
Force

When a force acts on a fixed object it is deformed, i.e., stretched or squashed, twisted or bent. When a force acts on a free object it changes its velocity, i.e., it accelerates or decelerates. (Fig. 3.1.) We can summarize the possible effects of forces:

1. Acceleration
2. Deformation

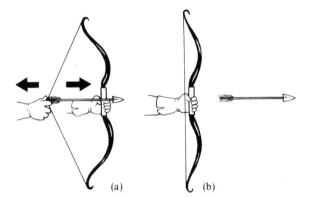

(a)   (b)

Fig. 3.1   (a) The hands apply balanced forces to the bow producing deformation. (b) When released the string exerts an unbalanced force on the arrow producing acceleration

Each of us has a first hand experience of force. No words can describe a force of tension as vividly as feeling an ear being pulled, or describe the force of compression as well as having a toe stood on. A good push at the right moment will send you flying and illustrate acceleration, while the force of collision with a hard object will cause you to suffer a rapid and probably painful deceleration. You are well equipped to understand forces so you can confidently trust your common sense.

Momentum ($m \times v$)

We start by describing a quantity called momentum which holds the key to understanding force.

The momentum of an object is defined as mass $\times$ velocity. The importance of the quantity is that it is conserved when objects collide, i.e., the total momentum before impact is equal to the total momentum after impact. This is called the law of conservation of momentum.

Momentum is a vector quantity because it has magnitude and direction. When adding linear momenta (momentums) those in one direction are reckoned positive and those in the other direction negative. You may understand momentum better if you follow how it applies in the following two common cases.

Consider what happens when two trucks of masses m_1 and m_2 collide. (Fig. 3.2.)

Total momentum before collision

$$= m_1 u_1 + m_2(-u_2)$$

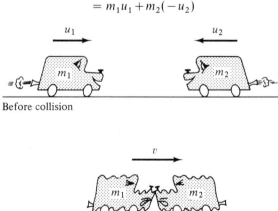

Before collision

After collision

Fig. 3.2   Momentum is conserved when objects collide

Suppose that they do not rebound on collision but have a common final velocity v in the direction of u_1.

Total momentum after collision

$$= m_1 v + m_2 v$$

By the law of conservation of momentum

$$m_1 u_1 + m_2 (-u_2) = m_1 v + m_2 v$$

The recoil of a rifle is explained by this conservation of momentum. (Fig. 3.3.) The initial momentum of the rifle and bullet is zero, so that, after firing, the nett momentum of the rifle and the bullet must also be zero.

$$m_1 v_1 + m_2 (-v_2) = 0$$
$$m_1 v_1 = m_2 v_2$$

i.e., the rifle has the same momentum as the bullet but in the opposite direction.

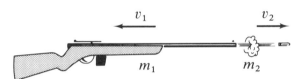

Fig. 3.3 At the instant of separation the rifle and the bullet have equal moments

Problem
A rifle weighing 3 kg fires a 2 g bullet with a muzzle velocity of 500 m/s. Calculate the velocity of recoil of the rifle.

Solution

$$m_1 v_1 = m_2 v_2$$

$$3 \times v_2 = \frac{2}{1000} \times 500$$

$$v_2 = \frac{2}{1000} \times \frac{500}{3} = 0.33 \text{ m/s}$$

velocity of recoil is 0.33 m/s.

Newton's Laws

Isaac Newton studied force and movement very carefully and came up with three laws of motion. They may seem simple but it needed his genius and at least one falling apple to recognize them.

Law 1 A body continues in a state of rest or uniform motion in a straight line unless acted upon by a resultant force.

Law 2 The rate of change of momentum ($m \times v$) of a body is proportional to the resultant force and takes place in the direction of the force.

Law 3 Every force produces an equal but opposed reactive force.

The first law means that if an object slows down, speeds up or deviates from a straight line, a resultant force must be acting on it.

The third law means that when one object presses on another then the other object presses right back with the same force. It is based on the fact that all forces arise from the interaction of pairs of bodies. Try standing with your toes touching a wall and then push on it and you get the sensation of it pushing back. (Fig. 3.4.)

The second law relates the resultant force to a change in the momentum $m \times v$.

Fig. 3.4 Action produces an equal and opposite reaction

Unit of Force: The Newton

From the second law of motion.

Force $\propto$ rate of change of momentum

i.e., Force $\propto$ rate of change of (mass $\times$ velocity)

Assuming that the mass does not change.

Force ∝ mass × rate of change of velocity,
but rate of change of velocity = acceleration

therefore

Force ∝ mass × acceleration

i.e., Force = k mass × acceleration

where k is a constant.

Now if we choose the unit of force to be of value 1 when the mass = 1 kg and the acceleration is 1 m/s², then $k = 1$ and we can write

Force = mass × acceleration
$$F = ma$$

This equation now defines the unit of force called the newton. **The newton is that force which gives a mass of 1 kg an acceleration of 1 m/s².**

Problem
What force produces an acceleration of 10 m/s² in a mass of 4 kg?

Solution
$$F = ma$$
$$= 4 \times 10 = 40 \text{ newtons}$$
$$\text{Force} = 40 \text{ N}$$

Problem
The force of gravity produces an acceleration of 9.81 m/s² towards the earth's surface. What is the force in newtons acting on a mass of 6 kg.

Solution
$$F = ma$$
$$= 6 \times 9.81 = 58.86 \text{ N}$$
$$\text{Force} = 58.86 \text{ N}$$

Pressure

Our sense of touch responds to pressure rather than force. A moderate force acting on a small area can be painful. (Fig. 3.5.) **Pressure is defined as the force acting perpendicularly per unit area.**

$$\text{Pressure} = \frac{\text{force} \quad \text{(in newtons)}}{\text{area} \quad \text{(in square metres)}}$$

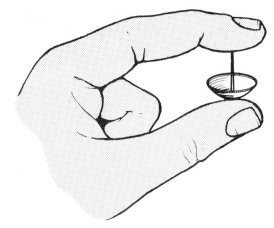

Fig. 3.5 Large and small pressures produced by the same force

The units of pressure are newtons per square metre (N/m²). Another name for the same unit is the Pascal.

$$1 \text{ N/m}^2 = 1 \text{ Pa}$$

Figure 3.6 shows how pressure depends on area. Cutting tools are given sharp edges (small areas) so that moderate forces apply very high pressures sufficient to part the work material.

Force = 2 × g = 2 × 9.81 = 19.62 N

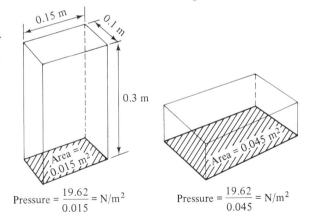

$$\text{Pressure} = \frac{19.62}{0.015} = \text{N/m}^2 \qquad \text{Pressure} = \frac{19.62}{0.045} = \text{N/m}^2$$

Fig. 3.6 Pressure exerted by a block of mass 2 kg resting on different faces

Problem
What is the pressure exerted by a force of 30 N on a chisel blade 20 mm wide and 0.1 mm thick.

Solution

$$\text{Pressure} = \frac{\text{force}}{\text{area}} = \frac{30}{0.20 \times 0.001}$$

$$= 150\,000 \text{ N/m}^2$$

Weight

Weight is a force. It is the force exerted on an object in the earth's gravitational field. We can determine the magnitude of the force on a mass m if we know its acceleration a when it falls freely.

$$F = ma$$

The acceleration at the earth's surface is denoted by the symbol g and has a standard value of 9.81 m/s² although the precise figure varies from place to place.

$$F = mg = m \times 9.81 \text{ newton}$$

Thus the force of the earth's gravity on a mass of 1 kg is approximately 10 N.

Do not confuse mass with weight. Mass is a measure of the amount of matter in an object. Mass determines the inertia of an object, i.e., how much it responds to force. Once the material is specified then the mass determines such things as its heat capacity or how much energy is released when it burns.

Weight on the other hand is the force exerted by a gravitational field and it varies as gravity changes. On the moon the gravitational force on a kilogram of matter has only one sixth of its value on earth, whereas other quantities depending on mass such as inertia remain the same.

Strength of Materials

A force applied to a fixed object causes it to stretch, to compress, to shear or to twist. (Fig. 3.7.) Just how much it deforms depends on the material, its shape and size, and the strength of the applied force. In many real situations the deformation is not apparent but the nature of the stress can be deduced. (Fig. 3.8.)

An elastic deformation is one in which the original shape is restored when the load is removed. Beyond a certain point the material takes

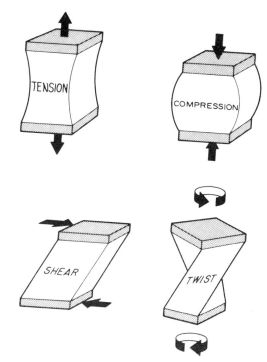

Fig. 3.7 Distortion produced by forces

a permanent set and does not fully recover and this point is called the elastic limit. Brittle materials do not deform much before they snap while plastic materials go on changing their shape as the load increases.

A rubber band is what most of us think of as elastic and so it is, but we must widen our view of elastic to include things like steel ball bearings where the deformation does not show but the recovery after load is pretty good.

When a gradually increasing load is applied to a suspended wire it progressively stretches until eventually it breaks. This test to destruction reveals the elastic properties of materials which can be deduced from the graph of load against extension. (Fig. 3.9.)

The straight line from O indicates that the extension is proportional to the load.

EXTENSION ∝ LOAD

This is Hooke's Law. (He invented the universal joint and helped to rebuild London after the great fire.) The straight line section of the graph is called the Hooke's Law region. Loads beyond the

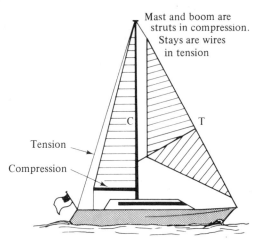

Mast and boom are struts in compression. Stays are wires in tension

Tension

Compression

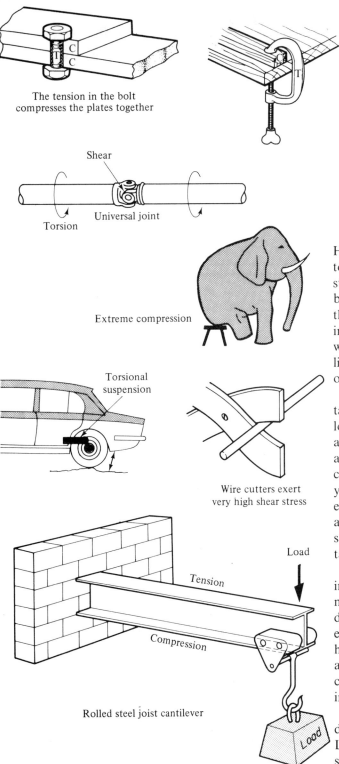

The tension in the bolt compresses the plates together

Shear

Universal joint

Torsion

Extreme compression

Torsional suspension

Wire cutters exert very high shear stress

Load

Tension

Compression

Rolled steel joist cantilever

Load

Fig. 3.8   Examples of various types of stress

Hooke's Law region are not directly proportional to the extension, and the graph deviates from a straight line. Eventually a point E is reached beyond which removing the load does not restore the wire to its original length. The dotted line indicates the set OS that would occur if the load were removed. The point E defines the elastic limit. It is usually (but not always) near to the end of the straight line section.

At Y the yield point, the curve becomes horizontal, i.e., the wire yields without any increase in load. In this region internal slipping of layers of atoms takes place. The crystal grains change shape as the specimen is drawn out and the metal becomes stronger and 'work hardens'. Thus after yielding the tension may increase further as plastic extension progresses. Eventually the wire develops a 'neck' at its weakest point and yields completely so that it breaks. The maximum tension it sustains, T, is the Ultimate tensile strength.

In practice to show these features in detail we impose a gradual extension on the specimen and measure the resulting tension by tensometer. For a ductile or plastic material the graph beyond the elastic limit is extended. For a brittle material however breaking occurs near to the elastic limit and no other features are measurable. For example cast iron which is brittle breaks before its length increases by 1%. (Fig. 3.10.)

The graph of Fig. 3.9 is characteristic of a ductile metal. Not all materials obey Hooke's Law. Soft rubber has little or no straight line section to its graph although it recovers well after the extensions. (Fig. 3.11.)

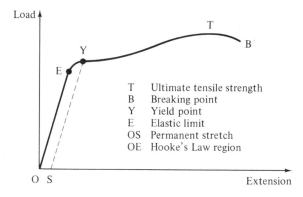

Fig. 3.9 **Load extension graph for a ductile metal (copper)**

T	Ultimate tensile strength
B	Breaking point
Y	Yield point
E	Elastic limit
OS	Permanent stretch
OE	Hooke's Law region

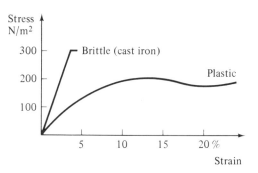

Fig. 3.10 **Stress and strain graphs for a brittle material and a plastic material**

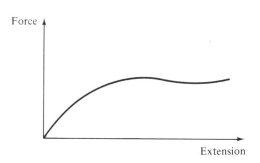

Fig. 3.11 **Many non-metallic substances are elastic but do not obey Hooke's Law (no straight line)**

Stress and Strain

The graph of load against extension depends on the shape and size of the specimen, which is incidental to the nature of the material. We can make a graph show only the elastic properties of the material as follows. We standardize the force by relating it to the cross-section of the specimen.

$$\text{Stress} = \frac{\text{load}}{\text{cross-sectional area}} \quad (\text{N/m}^2)$$

Note that the cross-sectional area is at right angles (normal) to the force. (Fig. 3.12.) We standardize the distortion by taking the fractional change in length.

$$\text{Strain} = \frac{\text{change in length}}{\text{original length}}$$

Strain does not have any dimensions because it is the ratio of two lengths and it is often quoted as a percentage.

Hooke's Law is frequently expressed as stress and is proportional to strain, i.e.,

$$\text{Stress} \propto \text{Strain}$$

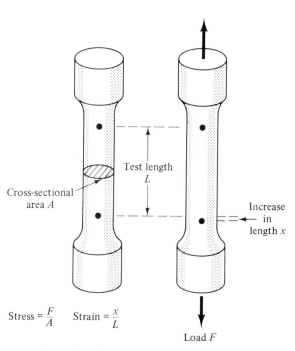

$$\text{Stress} = \frac{F}{A} \qquad \text{Strain} = \frac{x}{L}$$

Fig. 3.12 **The meaning of stress and strain**

27

Modulus of Elasticity

The slope of the stress–strain graph below the elastic limit is characteristic of the test material. It is called the Young modulus of elasticity E

$$E = \frac{\text{stress}}{\text{strain}} = \frac{\text{force per unit area}}{\text{fractional change in length}}$$

Young's modulus is a measure of the stiffness of the material. Some materials have the same value of E in tension as they have in compression but others have different values. (Table 3.1.) Brick and

Table 3.1   Typical Values of Elastic Constants
(some variation according to purity, etc.)

Material	Young modulus $\times 10^9$ N/m^2	Ultimate tensile stress $\times 10^6$ N/m^2
Aluminium	70	120
Copper	130	300
Mild steel	210	450
Cast iron	150	160
Concrete	30	4
Glass	70	70
Lead	16	15
Wood (oak)	70	90
Nylon	2	85

concrete are strong in compression but weak in tension. Hence concrete is reinforced with steel rods or girders when it is used structurally. (Fig. 3.13.)

Problem

A force of 50 000 N produces an extension of 2.1 mm in a brass rod of length 100 mm and diameter 15.2 mm. Calculate the stress, the strain and the value of the Young modulus for the rod.

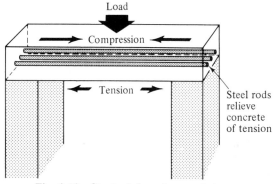

Fig. 3.13   Steel reinforced concrete beam

Solution

$$\text{Stress} = \frac{\text{force}}{\text{area}} = \frac{50\,000}{\pi \times (7.6/10^3)^2} = 2.76 \times 10^8 \text{ N/m}^2$$

$$\text{Strain} = \frac{\text{change in length}}{\text{original length}} = \frac{2.1}{100} = 0.021$$

$$E = \frac{\text{stress}}{\text{strain}} = \frac{2.76 \times 10^8}{0.021} = 1.3 \times 10^{10} \text{ N/m}^2$$

Young modulus $= 1.3 \times 10^{10}$ N/m^2

Shear

The ability of a material to withstand twist or shear is as important as its performance under tension or compression. (Fig. 3.14.) The equivalent of the Young Modulus for shear is the shear modulus

$$\text{Shear modulus} = \frac{\text{shear stress}}{\text{shear strain}}$$

Referring to Fig. 3.14

$$\text{Shear modulus} = \frac{\text{force/area}}{a/b}$$

Note that the area is parallel to the force.

$$\text{Shear modulus} = \frac{\text{Shear stress}}{\text{Shear strain}}$$

$$= \frac{F/A}{a/b} \text{ (N/m}^2)$$

Fig. 3.14   Shear modulus

Elastic properties are very dependent on temperature. For example glass is brittle when cold but can be drawn out to form the finest fibres when hot. Rapid cooling (quenching) at the final stage of production results in a harder and less ductile material.

The effect of a force on a material depends on whether the force is applied slowly or suddenly. Gradual application of the force allows the atoms to line themselves up and 'flow' into the new shape smoothly. A hammer blow applying the same force may cause a fracture because the atoms form a log jam and lock together.

Ductility and Malleability

Ductile materials can be drawn or compressed into very extended shapes without fracture under the action of a gradually applied force. Many materials become ductile when heated to near their melting points so we usually reserve the description ductile for those materials which can be drawn out when cold. For example copper, gold, silver and platinum can be cold drawn into very fine wires.

A malleable material can be hammered or rolled into extended shapes without cracking. Malleability particularly refers to the compressive force suddenly applied. For example gold leaf can be hammered out to a thickness of 0.001 mm. Other materials which are malleable when mildly heated are tin, aluminium and mild steel, hence their use in pans, foils and containers.

As you might expect ductile materials are also malleable but a material may be particularly outstanding in one or other property.

Brittleness

Brick, concrete, cast iron and glass are examples of brittle materials. They may be strong, they may be elastic within the limit, but when they break, they do so completely and without warning.

Brittleness is often more apparent in tension than in compression. Brunel built Paddington Station with cast iron pillars supporting a vast cathedral-like roof. Because the pillars were brittle the structure was thought to be unsafe and so they were later replaced with steel ones. This was just as well for a doodlebug bomb hit the station during the war. It did only minor damage but the story may well have been different with the original pillars.

Toughness

A tough material can absorb a lot of energy before it breaks. It is not brittle or too ductile but fights over every millimetre of its extension before breaking. The energy absorbed in breaking is represented by the area under the force extension curve so for tough materials this area is large. (Fig. 3.15.)

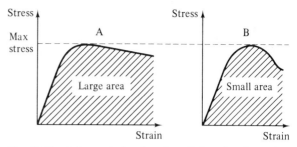

Fig. 3.15 A is much tougher than B but they have the same strength

Resilience

If a material can absorb the energy without exceeding the elastic limit then we say it is resilient. Wrought iron is tough and nylon is both tough and resilient.

Hardness

Hardness is the ability of a material to resist abrasion and indentation. Hardness is related to the elastic properties of a material but not in any simple way so that separate indentation tests are made. (Fig. 3.16.) It is an important property in many applications from cutting tools to floor coverings.

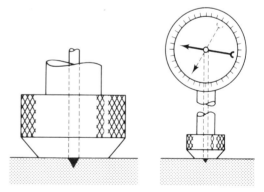

Fig. 3.16 Hardness testing. The indentation of the spring-loaded diamond tip is recorded on the gauge

Friction

The nearest you ever approach to experiencing a frictionless world is probably when you first attempt ice skating. If you have tried skating you will know that there are real problems of control.

(Fig. 3.17.) The same problem arises in hovercraft where friction has been reduced to a very low figure and accurate steerage is difficult. (Fig. 3.18.)

On the other hand excessive friction is the enemy of all moving machine parts. It absorbs energy and it causes wear. (Fig. 3.19.)

So what is friction precisely? It is the force acting parallel to surfaces in contact which tends to oppose their relative motion. (Fig. 3.20.) It is caused by the minute humps and hollows which make up even the smoothest surface. These lock into each other at rest and they must be forced apart or deformed locally for movement to occur. (Fig. 3.21.) When a force is exerted parallel to the surface an equal and opposite force of friction is set up. If the applied force is increased gradually the frictional force eventually reaches a limit and

Fig. 3.17   Lack of friction can give control problems

Fig. 3.18   Hovercraft do not steer precisely
(*Courtesy of Hoverlloyd Ltd*)

30

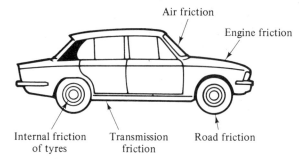

Fig. 3.19   Energy losses due to friction

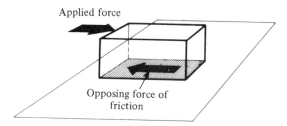

Fig. 3.20   The force of friction acts parallel to the surfaces so as to oppose motion

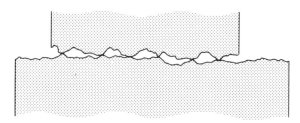

Fig. 3.21   Dry friction. The surfaces contact in only a few high pressure places

beyond this movement occurs in the direction of the applied force. A good deal of experimental evidence exists about this limiting force of friction. It indicates some gradual rules which are sometimes called the laws of friction but the evidence is not really precise enough to make them worthy of the title.

This picture of the surfaces in contact allows us to explain some of the experimental evidence.

1. Frictional force ∝ force acting normally to the surfaces.

This is to be expected since the greater the normal force the more difficult it is for the humps on the surfaces to push past each other.

2. Frictional force is independent of the area of contact.

Since the surfaces only really make contact in isolated places these cannot be expected to bear any relation to the apparent area.

3. The frictional force to start motion is greater than that required to maintain it.

When the surfaces are static the humps and hollows settle into each other.

4. Frictional force is independent of the velocity.

Not all the evidence supports this generalization and considering the argument used in 3 above, one might expect the friction to diminish at higher velocities. This effect must be counteracted by the inertia of the humps in being pushed aside and so the force remains approximately constant.

Coefficient of Friction

The ratio of the limiting frictional force to the force pressing the surfaces together is called the coefficient of friction μ. (Fig. 3.22.)

$$\mu = \frac{\text{limiting frictional force}}{\text{reactive force normal to surfaces}}$$

$$\mu = \frac{F}{R}$$

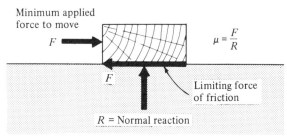

Fig. 3.22   The coefficient of friction expressed in terms of the force which causes movement

For particular surfaces μ has two values:

1. a static coefficient based on a start from rest
2. a dynamic coefficient based on the maintenance of motion

The dynamic coefficient is always smaller than the static coefficient .

Table 3.2 **Typical Dynamic Coefficients of Limiting Friction for Dry Surfaces**

	μ
Steel–steel	0.25
Steel–steel lubricated	0.03
Steel–cast iron	0.20
Rubber on asphalt	0.65
Rubber on concrete	0.75
Wood on wood	0.4
Steel on ice	0.04

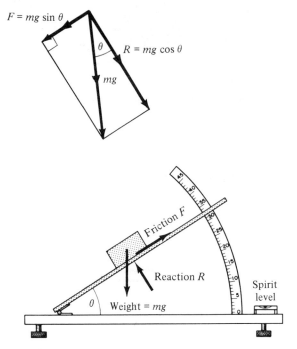

Fig. 3.23   Measuring friction by inclined plane

Problem

What force is required to move a block of mass 8 kg along a horizontal plane if the coefficient of static friction is 0.4 and $g = 9.8 \text{ m/s}^2$.

Solution

The normal reaction R is the weight of the block mg

$$R = mg = 8 \times 9.81$$

Coefficient of friction

$$\mu = F/R$$

The force required

$$F = \mu R = 0.4 \times 8 \times 9.81$$
$$\text{Force} = 31.4 \text{ N}$$

Problem

A force of 50 N is applied horizontally to an object of 12 kg resting on a horizontal surface. The coefficient of dynamic friction is 0.3. Calculate the acceleration of the object.

Solution

Normal reaction $R = mg = 12 \times 9.81$ N
Friction force $= \mu R = 0.3 \times 12 \times 9.81$
$$= 35.3 \text{ N}$$

Thus resultant force
$$= 50 - 35.3 = 14.7 \text{ N}$$

This produces an acceleration a

$$a = F/m = \frac{14.7}{12} = 1.2 \text{ m/s}^2$$

The acceleration is 1.23 m/s^2.

One method of measuring μ involves the tilting of a plane until sliding just occurs. (Fig. 3.23.)

The component of the weight of the object acting along the plane provides the slipping force and the component at right angles to the plane provides the normal reaction.

(See Chapter 4 for an explanation of component forces.)

$$\mu = \frac{F}{R} = \frac{mg \sin \theta}{mg \cos \theta} = \frac{\sin \theta}{\cos \theta} = \tan \theta$$

Thus the tangent of that angle of the plane which is just sufficient to start motion gives the coefficient of static friction directly. Similarly the dynamic coefficient is the tangent of the angle which maintains a constant velocity of the object once started by a helping push.

Friction is dependent on the materials in contact but it is also very dependent on surface finish. Finely sanded wood surfaces present less friction than rough sawn ones. Further, the friction is strongly affected by other substances between the surfaces. Grit on icy roads gives tyres a better grip while graphite between surfaces reduces the friction drastically. A liquid lubricant between the surfaces keeps them apart and replaces solid friction by fluid flow. (Fig. 3.24.)

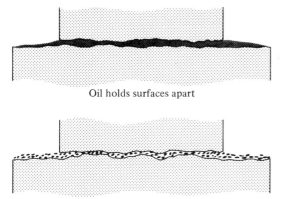

Oil holds surfaces apart

Solid graphite particles slide easily and separate the surfaces

Fig. 3.24 Liquid and solid lubricants

When a liquid flows its molecules are dragged across each other. The liquid offers a resistance to the flow called the viscosity which is equivalent to friction in solids. A liquid of high viscosity like gear oil can sustain its lubricating properties when subject to high pressures. Viscous liquids however offer a high resistance to motion and they were difficult to transport through pipes.

Highly mobile liquids which flow easily such as ether or petrol have low viscosities. For a particular lubrication task the viscosity must be high enough not to be squeezed out by the pressure and low enough not to give too much resistance. The viscosity of a liquid is characterized by a coefficient of viscosity η. Rates of flow through tubes are directly proportional to viscosity and comparisons are made by timing the flow of the liquid through standard tubes or orifices. (Fig. 3.25.)

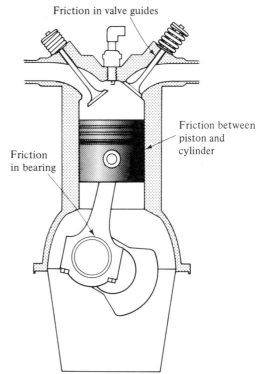

Fig. 3.26 Frictional losses

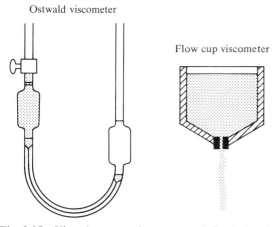

Ostwald viscometer

Flow cup viscometer

Fig. 3.25 Viscosity comparisons are made by timing the flow of liquid through tubes or holes

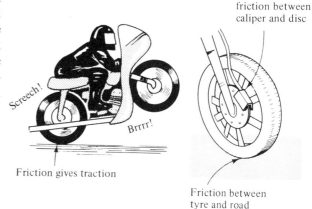

Braking friction between caliper and disc

Friction gives traction

Friction between tyre and road

Belt drive

Fig. 3.27 Examples of uses of friction

33

Water is a poor lubricant for metal surfaces but a good lubricant for rubber which is not always an advantage.

Friction of one kind or another absorbs most of the energy supplied to machines. It is converted to heat and dissipated to the surroundings. (Fig. 3.26.) Friction also finds uses in controlling and transmitting energy. (Fig. 3.27.) Roller or ball bearings can reduce the friction substantially. (Fig. 3.28.)

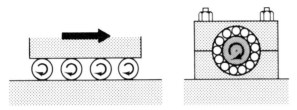

Fig. 3.28 Rolling friction is much less than sliding friction

Questions

1. How could you compare the masses of two objects without weighing them?

2. If you push a car, it pushes back with an equal and opposite force. Why then does it move?

3. What are the advantages of specifying stresses and strains rather than loads and extensions?

4. A rocket which burns fuel at a constant rate is fired vertically upwards into the atmosphere. Suggest reasons for the acceleration of the rocket increasing as it rises.

5. Explain how a nylon bracket could be tougher but weaker than a cast iron bracket.

Multiple Choice

1. The unit of force is
   (a) joule, (b) pascal, (c) newton, (d) kilogram.

2. The weight of an object is
   (a) a force, (b) an acceleration, (c) a mass,
   (d) a gravitational acceleration.

3. The acceleration produced by a force on an object is proportional to

   (a) $\dfrac{1}{\text{mass}}$, (b) mass, (c) $\dfrac{1}{\text{mass}^2}$, (d) mass^2.

4. The force of gravity on a kilogram mass is approximately
   (a) 1 N, (b) 10 N, (c) 100 N, (d) 1000 N.

5. Pressure is given by
   (a) force × area, (b) force × density,

   (c) $\dfrac{\text{force}}{\text{area}}$, (d) $\dfrac{\text{force}}{\text{mass}}$.

6. Hooke's Law states that load is proportional to
   (a) stress, (b) elastic limit, (c) extension,
   (d) ultimate tensile strength.

7. The elastic limit of a stretched wire is marked by
   (a) breaking, (b) permanent stretch, (c) necking,
   (d) maximum extension.

8. The quantity load/cross-sectional area is called
   (a) stress, (b) strain, (c) extension,
   (d) Young modulus.

9. The Young modulus is defined as
   (a) force/area, (b) stress/strain, (c) breaking stress,
   (d) fractional extension.

10. A substance which breaks suddenly without warning is described as
    (a) ductile, (b) tough, (d) strong, (d) brittle.

11. Which is *not* true of limiting frictional forces?
    (a) opposes motion,
    (b) proportional to normal reaction,
    (c) proportional to area,
    (d) proportional to coefficient of friction.

12. Viscous forces increase with
    (a) speed of flow, (b) density of fluid,
    (c) mass of fluid, (d) pressure.

Problems

1. A force of 20 N is applied to a mass of 5 kg. Calculate the acceleration produced. [4 m/s^2]

2. What force is required to give (a) 50 kg an acceleration of 0.4 m/s^2? (b) 2 kg an acceleration of 6 m/s? [20 N, 12 N]

3. A motor cycle decelerates from 60 km/h to rest in 3.0 s. What average force does a rider of mass 80 kg experience? [444 N]

4. What force is required (extra to resistive forces) to make a car of mass 700 kg reach 90 km/h in 8.0 s? [2.2 kN]

5. A force of 20 N acts for 8.0 s on a mass of 60 kg. What is the (a) acceleration, (b) final speed, (c) average speed, (d) distance covered during the 8.0 s? [0.33 m/s^2; 2.7 m/s; 1.35 m/s; 10.7 m]

6. A rectangular box measures 2.0 m × 1.5 m × 0.5 m and has a mass of 250 kg. Calculate the force of gravity in newtons acting on the box and the average pressure when it rests horizontally on each of its sides. (Take $g = 9.8$ m/s^2.) [2450 N; 817 N/m^2; 3270 N/m^2; 2450 N/m^2]

7. A woman weighs 400 N and has fashion shoes with approximately 12 mm square heels, and also walking shoes with heels of 50 mm by 50 mm. Calculate the pressure in each case when her weight is taken on one heel. [2.8 MPa; 160 kPa]

8. Estimate the force exerted by the atmosphere on a television screen measuring 400 mm by 500 mm. (Atmospheric pressure is 1.0×10^5 Pa.) [20 kN]

9. If a force of 500 N extends a wire by 2.2 mm, what force would extend the wire by 3.0 mm assuming Hooke's Law to be obeyed? [682 N]

10. Draw a load extension graph from the following data:

load (N)	500	1000	1500	2000	2500	3000
extensions (mm)	0.15	0.30	0.45	0.60	0.76	0.96

What is the extension produced by a load of 1650 N? Estimate the load at which Hooke's Law ceases to apply. [0.495 mm; just above 2000 N]

11. A wire 1.00 m long extends to 1.05 metres when a load of 2.4 kg is applied. What would be its length when 6.0 kg is applied. [1.125 m]

12. A load of 40 000 N produces an extension of 2.4 mm in a metal rod of square section of side 12 mm and length 1.5 m. Calculate the stress, the strain and the value of Young Modulus for the metal. [2.78×10^8 N/m^2; 0.0016; 1.74×10^{11} N/m^2]

13. An iron casting has a mass of 60 kg and slides on a horizontal metal surface. It requires a force of 240 N to start it moving but only 170 N to keep it going. Calculate (a) the normal reaction between the casting and the metal surface, (b) the coefficient of static friction, (c) the coefficient of dynamic friction ($g = 10$ m/s^2.) [600 N; 0.4; 0.28]

14. A racing car must be able to decelerate at 5 m/s^2 (50 m/s to rest in 10 s). It has a mass of 1 tonne (1000 kg). What is the minimum coefficient of friction tolerable between the car and the road? [0.5]

15. A new type of ski for walking up snow slopes has been proposed by a Swedish engineer. They have nylon hairs on the bottom which allow the ski to slide forwards only. The forward coefficient of friction is almost zero. The reverse coefficient is 0.9. What is the maximum slope up which you could walk in them? [42°]

Multiple Choice Answers

1(c), 2(a), 3(a), 4(b), 5(c), 6(c), 7(b), 8(a), 9(b), 10(d), 11(c), 12(a).

4.
Statics

Most of the things about us are stationary. They are balanced, wedged, fastened or otherwise supported. Our sanity, even our survival depends on it. Moving things pose a threat and therefore must command our attention against a background of still objects. But the stillness disguises an array of forces which interact to reach a balance so holding the objects at rest.

Statics is the study of forces in equilibrium. When we fail to understand and regulate the statics of our structures then they move or collapse and we fail in a way that can be tragic or just plain inconvenient.

Centre of Gravity

The weight of an object is one of the commonest and most tangible forces we have to deal with. The position in the object where this force effectively acts is called the centre of gravity. When you put a ruler, a book, plate down near the edge of a table you make sure the centre of it is inside the edge and you demonstrate that you know about centres of gravity and where they are situated. (Fig. 4.1.) Another name for the centre of gravity is the centre of mass. It is an important consideration in assessing the road holding of vehicles. (Fig. 4.2.) Even when there is no danger of a car rolling on a bend the position of the centre of gravity determines how the weight is spread between inside and outside wheels and hence the skidding and cornering characteristics. For complex shapes the

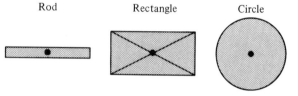

Rod   Rectangle   Circle

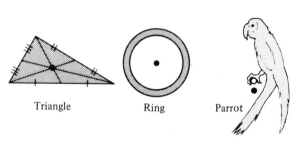

Triangle   Ring   Parrot

Fig. 4.1 Position of the centre of gravity of simple shapes

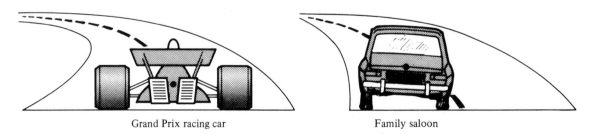

Grand Prix racing car   Family saloon

Fig. 4.2 Position of centre of mass is an important factor in automobile design

centre of gravity is obtained by balancing or tilt tests. (Fig. 4.3.)

We can find the geographical centre of a map by suspending its cut out shape on card from two positions. The intersection of the verticals drawn by suspending a plumb line gives the centre of gravity of the shape. This might be the best place to site a radio transmitter.

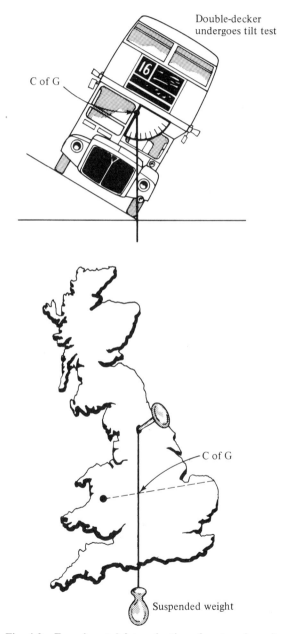

Fig. 4.3   Experimental determination of centre of gravity

Moments

An object free to pivot will rotate when a force is applied which does not pass through the pivot. The turning effect is proportional to:

1. The magnitude of the force
2. The perpendicular distance from the force to the pivot.

The product of these two quantities is called the moment of the force about the pivot. (Fig. 4.4.)

Moment = force × perpendicular distance

Figure 4.5 illustrates that it is the perpendicular distance that counts not just the point of application. A force which acts through the pivot has a zero perpendicular distance and therefore a zero moment. (Fig. 4.6.)

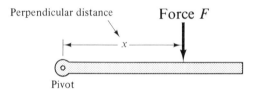

$$\text{Moment} = Fx$$

Fig. 4.4   Moment = force × perpendicular distance

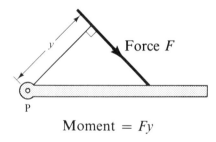

$$\text{Moment} = Fy$$

Fig. 4.5   Moment is proportional to the perpendicular distance

Fig. 4.6   Forces which act through the pivot do not exert a turning effect and have no moment

We can express the turning effect of a force (torque) about any point we care to choose by calculating the moment of the force about that point. (Fig. 4.7.)

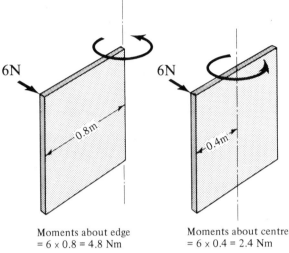

Moments about edge
= 6 x 0.8 = 4.8 Nm

Moments about centre
= 6 x 0.4 = 2.4 Nm

Fig. 4.7   Moments about any point

Where several forces act on an object the resulting moment about any point is obtained by summing the separate moments of each of the forces. When making this sum consider whether each force exerts a clockwise or anticlockwise turning effect about the chosen point. Clockwise moments are counted positive, and anticlockwise moments are counted negative. (Fig. 4.8.)

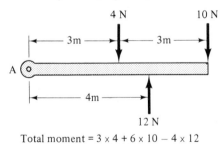

Total moment = 3 x 4 + 6 x 10 − 4 x 12
Moment about A = 24 Nm

Fig. 4.8   Calculation of the resultant moment about A

When an object is in equilibrium the sum of the moments about any point must be zero (or else it would turn about that point). This is the Principle of Moments. Stated another way the clockwise moments must equal the anticlockwise moments about any point, i.e.,

Clockwise moments = anticlockwise moments

38

Problem

A uniform beam of weight 400 N and 6 m long is supported horizontally at its ends. Calculate the force on the supports when a load of 250 N is placed 1 m from one end. (Fig. 4.9.)

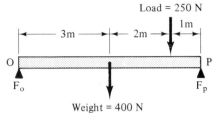

Fig. 4.9   Beam in equilibrium

Solution

Since the beam is in equilibrium

$$CM = ACM$$

To find F_p take moments about O

$$400 \times 3 + 250 \times 5 = F_p \times 6$$

$$F_p = \frac{400 \times 3 + 250 \times 5}{6} = \frac{2450}{6}$$

Force on p = 408.3 N

To find F_o take moments about P

$$F_o \times 6 = 400 \times 3 + 250 \times 1$$

$$F_o = \frac{400 \times 3 + 250 \times 1}{6} = \frac{1450}{6}$$

Force on O = 241.7 N

Equilibrium

We are now in a position to summarize the conditions that must be satisfied if a body is in equilibrium:

1. Moments about any point must sum to zero
2. Forces in any direction must sum to zero

When 1 is satisfied but not 2 then the body moves in the direction of the resultant force without rotation. When 2 is satisfied but not 1, the body rotates but stays where it is. When neither condition is satisfied it spins away into outer space!

Problem

Confirm that the beam AB is in equilibrium. (Fig. 4.10.)

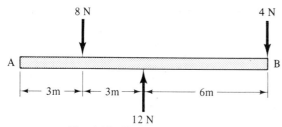

Fig. 4.10   Beam in equilibrium

Solution

For no rotation moments must sum to zero about any point, i.e.,

clockwise moments = anticlockwise moments

take moments about A

$$8 \times 30 + 4 \times 120 = 12 \times 60$$
$$720 = 720$$

Therefore the beam does not rotate.

To check that there is no movement
Force acting vertically down $= 8 + 4 - 12 = 0$
Hence the beam is in equilibrium.

Problem

A load of 2000 N is concentrated at a point 2.5 metres from the end of a 6 m beam of weight 800 N. Calculate the force on the supports at 1 m from each end. (Fig. 4.11.)

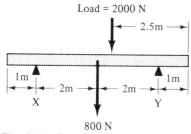

Fig. 4.11   Forces on a supported beam

Solution

By taking moments about a support X we eliminate the need to know the force at this point and we can write an equation with only one unknown quantity.

$$CM = ACM$$

$$800 \times 2 + 2000 \times 2.5 = F \times 4$$

$$F = \frac{1600 + 5000}{4} = 1650 \text{ N}$$

To find the load on the support X we could either take moments about Y or we can use the second condition for equilibrium which is easier.

Total upward force = Total downward force
$$1650 + F_x = 2000 + 800$$
$$F_x = 2000 + 800 - 1650$$
$$\text{Force at } X = 1150 \text{ N}$$

Problem

A uniform beam 5 m long of mass 300 kg is supported at a point 0.5 m from its mid-point. A mass of 1200 kg is applied at the end nearest to the support. What mass m placed at the other end would produce equilibrium? (Fig. 4.12.)

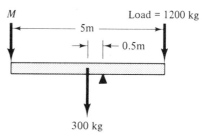

Fig. 4.12   Beam balanced by masses

Solution

The weight of the beam acts through its centre of gravity at its mid-point. To convert the masses into forces we use

$$F = \text{mass} \times \text{acceleration} = mg$$

Where g is the acceleration due to gravity.
Taking moments
$$CM = ACM$$
$$1200 \text{ g} \times 2 = mg \times 3 + 300 \text{ g} \times 0.5$$
divide through by g
$$1200 \times 2 = m \times 3 + 300 \times 0.5$$

$$m = \frac{1200 \times 2 - 300 \times 0.5}{3}$$

$$\text{mass} = 750 \text{ kg}$$

Note that where all the forces are expressed as the weights of masses we can omit the conversion to force units.

Stability of Equilibrium

An object may satisfy the condition for equilibrium and yet be in a precarious condition. A pencil balanced on its end demonstrates this well. Quite a small disturbance causes it to topple and we say it is unstable. In contrast, a book resting flat down is very stable and would have to be lifted to turn it over. (Fig. 4.13.)

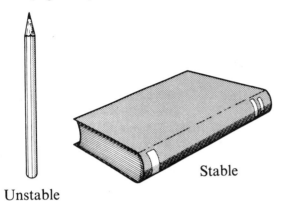

Unstable   Stable

Fig. 4.13   Objects in equilibrium

Just how stable a particular system is can be estimated by considering what happens to the centre of gravity when small movements occur. If after a small movement the centre of gravity moves downwards then the equilibrium is unstable. (Fig. 4.14.) If the centre of gravity continues to rise during quite large movements, the equilibrium is stable. Situations do exist where the centre of gravity moves neither up nor down and we have neutral equilibrium.

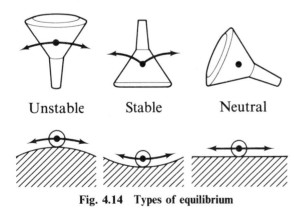

Unstable   Stable   Neutral

Fig. 4.14   Types of equilibrium

Scalars and Vectors

We divide physical quantities into two major categories. Those which have magnitude and direction we call vector quantities, e.g., force, velocity, ac-

celeration. Those which have only magnitude we call scalar quantities, e.g., mass energy volume.

The reason why we divide quantities into scalars and vectors is because of the different ways they are added together. Scalar quantities are very easy to combine since they can be simply added arithmetically, e.g., adding masses.

$$2\,\text{kg} + 4\,\text{kg} = 6\,\text{kg}$$

Vectors on the other hand must be added geometrically, i.e., by scale drawings which allow for their size and their direction. Each vector quantity can be represented by a straight line. The direction of the line is parallel to the vector. The length of the line is proportional to the magnitude of the vector. (Fig. 4.15.)

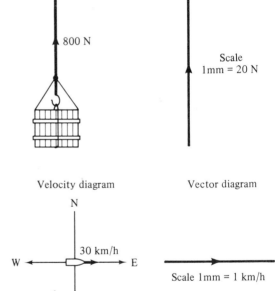

Fig. 4.15   Representing vector quantities by scale drawing

Parallelogram of Vectors

To find the resultant of two vectors we use a simple construction.

Take the case of two forces acting on an object. (Fig. 4.16(a).) We represent these forces by straight lines drawn to scale. (Fig. 4.16(b).) Next

(a) Force diagram

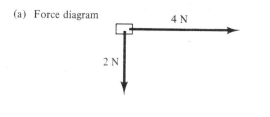

(b) Vector diagram

Scale 1 N = 10mm

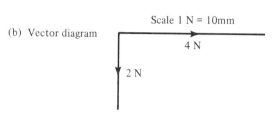

(c) Completed
    parallelogram

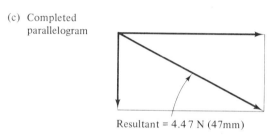

Resultant = 4.4 7 N (47mm)

Fig. 4.16   Stages in finding the resultant of two forces

Force diagram

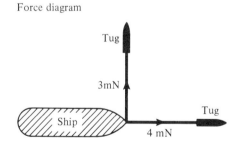

Vector diagram   Scale 5mm = 1 mN

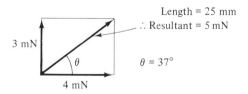

Fig. 4.17   Resultant of two forces

Velocity diagram   Vector diagram

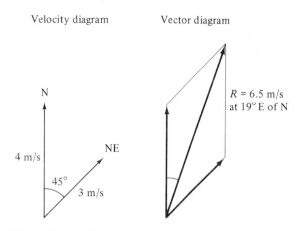

$R = 6.5$ m/s
at 19° E of N

Fig. 4.18   Resultant of velocities of 4 m/s N and 3 m/s NE

Velocity diagram   Vector diagram

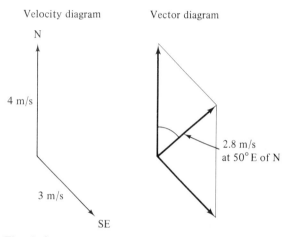

2.8 m/s
at 50° E of N

Fig. 4.19   Resultant of velocities of 4 m/s N and 3 m/s SE

we complete the parallelogram and draw the diagonal through the same point. The direction of this diagonal is that of the resultant force, and its length represents the magnitude of the resultant force on the same scale. (Fig. 4.16(c).) The method works equally well for other vector quantities and for situations where the angles between the components are greater or less than 90°. (Figs. 4.17, 4.18 and 4.19.)

Rectangular Components

Imagine an aeroplane diving at an angle of 30° to the horizontal at 80 m/s and that we wish to know at what velocity it is falling vertically. We can use the parallelogram of vectors in reverse to break up the plane's velocity into a vertical component and a horizontal component as follows (Fig. 4.20):

1. Draw the vector representing the velocity of the plane.
2. Using this as hypotenuse complete the rectangle with horizontal and vertical lines.
3. Deduce the components by measuring the sides of the rectangle.

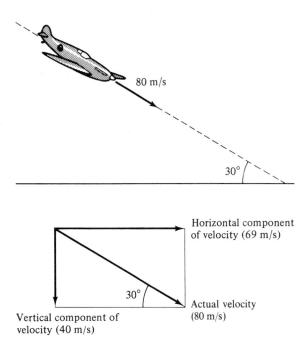

In the same way we can resolve the force on a sail into components acting parallel and at right angles to a yacht. (Fig. 4.21.) The keel of the yacht is designed to offer a high resistance to sideways motion and low resistance to forward motion.

The resolution of vectors into perpendicular components lends itself to easy expression by trigonometry. Thus in Fig. 4.21:

$$\frac{AB}{AC} = \cos \theta = \frac{\text{forward force}}{\text{total force}}$$

∴ Forward force = total force × cos θ

i.e.,

Resolved force = $F \cos \theta$

where θ is the angle between the force and the direction of the component.

As an example the problem of Fig. 4.20 could have been more easily solved by this method.

Velocity of plane = 80 m/s
Horizontal velocity = 80 cos 30° = 69.3 m/s
Vertical velocity = 80 cos 60° = 40.0 m/s

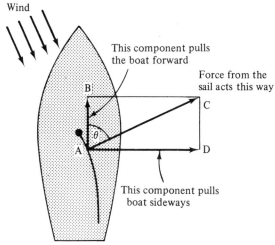

Fig. 4.21 Resolving the force acting on a sailing boat

The Triangle of Forces

Figure 4.22 shows how the parallelogram of forces gives the resultant of two forces acting at a point. If a third force acts at that point and is in equilibrium with the other two it must be equal and opposite to the resultant. We could have obtained the equilibrium force directly by simply constructing the triangle with the arrows all going the same way round. This principle is called the triangle of forces. **Three forces in equilibrium acting at a point can be represented in magnitude and direction by the sides of a triangle taken in order.** Note that for equilibrium the three forces must be in the same plane.

Problem
A load of 3000 N hangs from a wire which is pulled to one side by a force of 1200 N applied horizontally. Find the tension in the upper part of the wire.

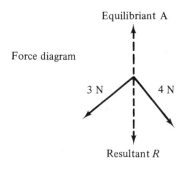

Force diagram

Equilibriant A

3 N   4 N

Resultant *R*

Parallelogram of force   Triangle of force

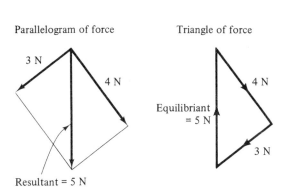

3 N

4 N   4 N

Equilibriant
= 5 N

3 N

Resultant = 5 N

Fig. 4.22 The resultant of two forces obtained by the parallelogram is exactly opposite to the equilibrium force obtained by the triangle of force

Solution

We use the triangle of forces for the point P where the three forces act. (Fig. 4.23.)

1. We choose a scale, e.g., 1 mm = 50 N
2. Draw Y parallel to PY 60 mm long
3. Draw X parallel to PX 24 mm long
4. Complete triangle and measure *F* (= 65 mm)
5. Use scale to convert to force

   Force = 65 × 50 = 3250 N
   Force = 3250 N at 68° to the horizontal

A further deduction we can make from the triangle of force is that where an object is in equilibrium under three non-parallel forces they must act through the same point. If they did not do so one force could not balance the resultant of the other two. This fact is useful in solving problems involving three forces where the direction of one of them is not obvious.

Force diagram   Vector diagram

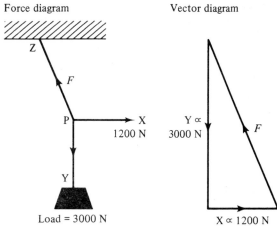

Z

F

P   X
         1200 N

Y

Load = 3000 N

Y ∝
3000 N

F

X ∝ 1200 N

Fig. 4.23   Use of the triangle of force

Problem

Find the tension in the control wire and the force exerted at the pivot of a heavy trap door in the position of Fig. 4.24.

Solution

Three forces act on the door: (1) the weight of the door acts down vertically through its mid point, (2) the tension in the wire which acts in the direction of the wire, and (3) force at the pivot. The tension and the weight act at point O where they intersect, hence the force on the pivot also acts through this point, i.e., along PO.

Draw AC vertically of length ∝ 1200
Draw CB parallel to PO
Draw BA parallel to RO

Measuring the lengths CB and BA on the same scale as AC we have

Force at pivot = 1220 N at 57° to the vertical
Tension in the wire = 700 N

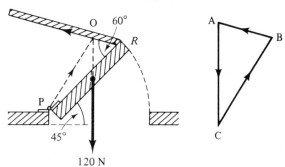

O   60°
              R
A
              B

P

45°

120 N

C

Fig. 4.24   Three forces in equilibrium act through the same point

Questions

1. Explain how it is possible to *raise* a load by exerting a force *downwards*.

2. Why are racing cars made low and wide?

3. Why do racing cars have very wide tyres?

4. Work and torque are both defined as force × distance. What is the difference between them?

5. Can you increase your power by the use of a lever?

Multiple Choice

1. Which is the vector quantity?
   (a) density, (b) mass, (c) time, (d) velocity.

2. Which is the scalar quantity?
   (a) energy, (b) acceleration, (c) displacement, (d) force.

3. Two equal and opposite forces F acting at the same point produce a resultant of
   (a) zero, (b) $\dfrac{F}{2}$, (c) F, (d) $2F$.

4. What values are possible for the resultant of two forces of 3 N and 4 N?
   (a) zero, (b) 0.5 N, (c) 6 N, (d) 8 N.

5. A fish exerts a tension of 9 N on a line attached to the end of a rod of length 3 m. What is the maximum torque it can exert on the rod?
   (a) $\frac{1}{3}$ N m, (b) 3 N m, (c) 9 N, (d) 27 N m.

6. A child of mass 30 kg sits 4.0 m from the pivot of a see-saw. How far from the pivot must a 40 kg child sit to balance?
   (a) 1.3 m, (b) 1.5 m, (c) 3.0 m, (d) 4.0 m.

7. When the moments acting on an object sum to zero but the forces do not, the object will
   (a) rotate on the spot, (b) move and rotate,
   (c) move with no rotation, (d) remain stationary.

8. Three forces are in equilibrium. If two of them act at a point at different angles then the third force
   (a) bisects them, (b) equals their sum,
   (c) acts through the same point,
   (d) is parallel to one force.

9. An egg in an egg cup has an equilibrium which is
   (a) stable, (b) unstable, (c) neutral.

10. Three parallel forces holding a body in equilibrium
    (a) sum to zero, (b) act through the same point,
    (c) exert no resultant moment,
    (d) act through the centre of mass.

Problems

1. A yacht sails due north at 4 km/h in a tide which travels due east at 3 km/h. What is the yacht's position after one hour? [5 km at 036.9°]

2. Two tow-lines, on opposite sides of a ship make angles of 30° and 40° with its direction of travel. The tensions in the lines are 20 kN and 15 kN respectively. Draw the vector diagram and determine the resultant forward force on the ship. [28.8 kN]

3. A weight of 4000 N hangs from the rope of a crane and is pulled aside by a horizontal force of 2000 N. Draw a scale diagram representing the weight and the force F and deduce (a) the direction of their resultant, (b) the tension in the rope of the crane. [26.6° to the vertical; 4500 N]

4. Two tractors each capable of exerting a force of 15 000 N are together towing a load. Calculate the maximum load they could move if their tow-lines are (a) parallel, (b) at 20°, (c) at 30°. [30 000 N; 28 200 N; 26 000 N]

5. A load of 5400 N is suspended from a rope. A horizontal force of 2000 N is exerted on the load. What then will be the tension in the rope and the angle it makes with the vertical? [5760 N at 20.3°]

6. A ladder weighing 200 N leans against a wall at 30° to the vertical. If the wall is smooth, so that the force it exerts on the ladder is horizontal, find the magnitude and direction of the force at the foot of the ladder. [208 N at 73.9° to horizontal]

7. A force of 200 N is exerted on the handle of a door 0.6 m from the hinge. If the force acts at right angles to the door, what is the moment it exerts? [120 N m]

8. A uniform beam 1.0 m long has a mass of 2.0 kg. It is supported at one end and at a point 250 mm from the other end. By taking moments about each of the supports in turn calculate the force exerted on each support and check that they sum to 20 N. Use $g = 10$ m/s². [0.7 N; 1.3 N]

9. A uniform rule 1.00 m long has a mass of 50 g. A mass of 100 g is attached to one end of the rule. Find the position of balance of the rule. [170 mm from the 100 g mass]

10. A crowbar 1.2 m long pivots at a point 80 mm from the end. A man applies his full weight of 750 N to the long end. What is the maximum load he could lift? [10 500 N]

11. A rod 1.00 m long is suspended at both ends by spring balances. The balances each register 1.0 N. What is the weight of the rod? Loads of 3.0 N and 4.0 N are hung from the rod at opposite ends. Calculate the readings on the spring balances. [2.0 N; 4.0 N; 5.0 N]

12. A uniform rod 1.00 m long has a mass of 0.05 kg hung from one end. The rod is balanced on a knife edge placed 10 cm from the end. Explain how the rod, with the mass attached, can rest in a horizontal position. Calculate the mass of the rod. [12.5 g]

13. A Tee-shape is made from two strips of aluminium measuring 60 mm × 20 mm. Calculate the position of the centre of gravity. [Centrally, 50 mm from base]

14. Calculate the position of the centre of gravity if the two strips of problem 13 are formed into an L-shape. [Inside edge 10 mm from corner]

Multiple Choice Answers

1(d), **2**(a), **3**(a), **4**(c), **5**(d), **6**(c), **7**(c), **8**(c), **9**(a), **10**(ac).

5.
Work, Energy and Power

We must distinguish clearly between the physical definition of work and our idea of activities which produce fatigue. Holding a mass quite still at arm's length, or pushing an object that will not budge, tires us very quickly, but strictly in neither case is work done. (Fig. 5.1.)

In science the term work has a precise meaning which is not related to mental or physical fatigue.

$$\text{Work} = \text{Force} \times \text{Distance}$$

The unit of work, the joule, is the work done when a force of one newton moves a distance of one metre.

The meaning of work is illustrated in Fig. 5.2. Note that movement perpendicular to the force does not contribute to the work done. Work is done only when a force moves in the direction in which it is applied. (Fig. 5.3.)

Problem

A load exerting a force of 120 N is carried up a flight of 15 stairs each 0.20 m high and 0.30 m from front to back. (Fig. 5.4.) How much work is done on the load?

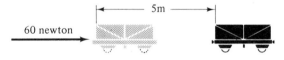

5m

60 newton

Work done = force × distance
= 60 × 5 = 300 joules
Work done = 300 J

Fig. 5.2   Calculation of work down in joules

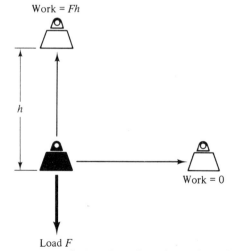

Work = Fh

h

Work = 0

Load F

Fig. 5.3   The work done by a force in moving a load is proportional to the distance moved parallel to the load

Fig. 5.1   If the force is not moved no work is done

Solution

$$\text{Work} = \text{Force} \times \text{Distance}$$

The horizontal measurement of the stair is not involved since work is only done against the force of gravity when the mass is moved vertically.

$$W = 120 \times (15 \times 0.2) = 90 \text{ J}$$
$$\text{Work done is } 90 \text{ joule.}$$

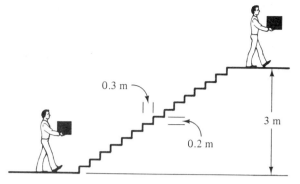

Energy

A person who performs a hard day's work is said to have expended energy. This use of the word energy coincides with its scientific meaning which defines energy as the capacity to do work.

Fig. 5.4 Work is done on a mass in moving it against the force of gravity. The horizontal movement does not contribute to the work

Plate 1   Power station
(*Courtesy of the CEGB*)

This capacity to do work may take many forms. Heat supplied to a steam engine results in the performance of work and therefore heat is a form of energy. The human body derives both its heat and its ability to work from the energy in food. This is described as chemical energy also possessed for example by fuels such as gas and oil. Electric motors and heaters demonstrate that electricity is a form of energy. There are many more kinds of energy such as light (solar panels) and atomic energy (nuclear reactors). They all have one thing in common, that is when suitably harnessed they can be made to do work.

It is important to note that energy is conserved, i.e., it does not appear from nowhere nor does it disappear without trace. It is only converted from one form into another. Some forms of energy are more spectacular than others. The chemical energy in a can of petrol is not too obvious until it is ignited. The electrical energy in a wire is unimpressive unless you touch it.

The forms of energy are all interchangeable. The energy in the can of petrol with a little ingenuity results in the motion of the car. An electric current will change the chemical composition of the plates of a car battery which will in turn release energy in the form of heat, light, or sound as required. Most forms of energy eventually result in heat which dissipates to the surroundings. In this case al-though the energy still exists it is no longer in a form in which we can readily use it. With a little thought you will be able to trace any sequence of energy conversion. (Fig. 5.5.)

Chemical Energy

It is obvious when a substance burns that energy is being released. This arises from the re-arrangement of the atoms which occurs in the chemical process of combustion. Where the chemicals react very rapidly the result may be an explosion. On the other hand chemical processes may release energy slowly and inconspicuously. Foods possess chemical energy which is converted by the body to heat and to movement. Batteries or cells give electrical energy and in so doing the chemicals of which they are made change their composition. Before all these energy producing reactions occur the substances possess chemical energy. When electrical energy is supplied to a rechargeable battery it is stored in the cell as chemical energy.

Kinetic Energy

A mass moving at speed possesses energy. It could do work using the inertia to push back a force before it is brought to a stop. This energy of motion is called kinetic energy. The ability of

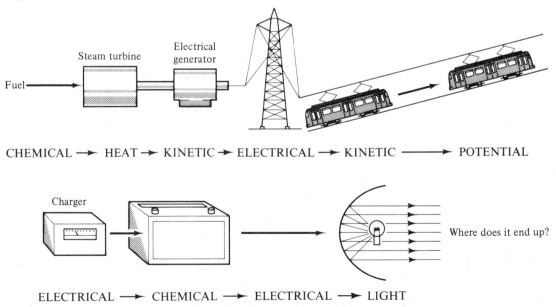

CHEMICAL → HEAT → KINETIC → ELECTRICAL → KINETIC → POTENTIAL

ELECTRICAL → CHEMICAL → ELECTRICAL → LIGHT

Fig. 5.5 Conversion of energy from one form to another

moving matter to do work is more apparent in the moving water driving turbines or simple water wheels, and in the wind driving wind vanes or yachts.

We can derive a simple expression for the kinetic energy of a mass m moving with a velocity v

$$\text{Kinetic energy} = \tfrac{1}{2}\text{ mass} \times \text{velocity}^2 = \tfrac{1}{2}mv^2$$

The energy will be expressed in joules when the mass m is in kilograms and the velocity v is in m/s.

Problem

Compare the kinetic energies of a car of mass 1000 kg travelling (a) at 12 m/s, (b) at 30 m/s, (approximately 30 mph and 70 mph.)

Solution

(a) $\text{KE} = \tfrac{1}{2}mv^2 = \tfrac{1}{2} \times 1000 \times 12^2 = 72\,000$ J
(b) $\text{KE} = \tfrac{1}{2}mv^2 = \tfrac{1}{2} \times 1000 \times 30^2 = 450\,000$ J

The car has approximately six times as much energy at 30 m/s than it has at 12 m/s.

Potential Energy

Winding a watch compresses a spring which later drives the watch mechanism. Energy is stored in the spring which is said to have potential energy, i.e., energy due to its position.

A mass raised above the ground has a potential energy because work could be done as it descends to a lower level. (Fig. 5.6.) The potential energy of

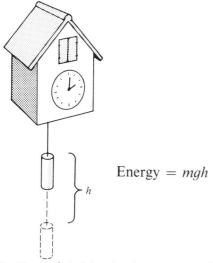

$$\text{Energy} = mgh$$

Fig. 5.6   The clock is driven by the energy stored in the raised mass as it descends under gravity

a mass m raised up a distance h is equal to the work done in moving it against the force of gravity. If g is the acceleration due to gravity:

$$\text{Force} = mg$$
$$\text{Potential energy} = \text{force} \times \text{distance}$$
$$= mgh \text{ joule}$$

Problem

How much energy is required to haul a lift of mass 600 kg from the bottom to the top of a sky scraper 200 m high? (Take $g = 9.8$ ms^{-2}.)

Solution

$$\text{Potential energy} = mgh$$
$$= 600 \times 9.8 \times 200$$
$$= 1\,176\,000 \text{ J}$$

Energy required is 1.176×10^6 J.

Nuclear Energy

Earlier science was based on two separate laws of conservation: the law of conservation of mass which states mass cannot be created or destroyed, and the law of conservation of energy—energy cannot be created or destroyed but only converted from one form to another. Nowadays, we know that neither law holds true in isolation.

In a nuclear explosion, a very large amount of energy is released, which is not apparently converted from any other form and, at the same time, a small amount of mass disappears. Both conservation laws are contradicted unless—and this idea opens up astonishing new fields of possibility—the small amount of mass and the large amount of energy are in some way related. We now believe that they are, in the sense that mass is a property of energy. Any form of energy has mass associated with it. Thus a given force accelerates an object less when it is hot than when it is cold. For example, a sphere of iron 2 m in diameter, which is glowing red hot, is about a tenth of a milligram more massive than when it is at room temperature. Einstein was the first to express the relation between mass and energy in an exact way by the equation:

$$\text{Energy} = \text{mass} \times (\text{velocity of light})^2$$
$$E = mc^2$$

E is given in joules whem m is expressed in kilo-grams and c is in m/s. The value of c is 3×10^8 m/s, so that $c^2 = 9 \times 10^{16}$.

This large constant of proportionality accounts for the apparently large difference between the energy and the mass associated with it.

Problem

How much energy is equivalent to a mass of a millionth part of a gram?

Solution

$$10^{-6}\,\text{g} = 10^{-9}\,\text{kg}$$
$$\text{We use } E = mc^2$$
$$= 10^{-9} \times (3 \times 10^8)^2$$
$$\text{Energy equivalent} = 9 \times 10^7 \text{ joule}$$

Radiation Energy

Light and heat radiation are just a part of the electromagnetic radiations which carry energy. X-rays, ultra-violet, radio and radar waves also carry energy and differ from light only in wavelength and frequency. Almost all the energy transfer in the universe is via electromagnetic waves. The stars, of which the sun is one, radiate vast amounts of energy into space. The earth receives only a minute fraction of the energy from the sun but it powers the winds and the waves and maintains the earth's temperature at a level suitable for life. Solar radiation panels will be used increasingly to generate energy as their production costs decrease and fuel costs rise.

Power

In these days when, as we say, 'time is money', we are concerned with not only doing a job, but doing it within a given budget and within a certain time. Thus, the rate at which energy can be supplied, the power, is as important to us as the total energy which is required. Power is defined as the rate of doing work.

$$P = \frac{\text{work}}{\text{time}}$$

Power may also be defined as the rate of transfer of energy. **The SI unit of power is the watt, a rate of working of 1 joule per second.** Where this unit is too small, the kilowatt (1000 watt) is used.

A man needs an average intake of 100 watt to maintain life but in short bursts as for example when running upstairs his mechanical power is as high as 2 kW. The sun radiates watts onto each square metre of the earth's surface, while a fan heater on its minimum setting usually gives a power of 1 kW. The old unit of power; the horse-power, is roughly the power that a horse can muster working hard for a short time.

Motor vehicle engines produce from about 30 to 300 kW (40 to 400 hp), aircraft engines several thousand kW, and some marine engines about 100 000 kW. Electric motors will develop 20 000 kW or more, but it is generally more convenient to use a number of small ones of low power than to use one large motor. (Fig. 5.7.)

Problem

A jet plane of mass 5000 kg can climb vertically at a rate of 200 km/h. What power is being expended? ($g = 9.81$ m/s.)

Solution

$$\text{Force} = mg = 5000 \times 9.81 \text{ N}$$
$$\text{Distance} = 200 \text{ km} = 200\,000 \text{ m}$$
$$\text{Time} = 50 \times 60 \text{ s}$$

$$\text{Power} = \frac{\text{work done}}{\text{time taken}} = \frac{\text{force} \times \text{distance}}{\text{time}}$$

$$= \frac{5000 \times 9.81 \times 200\,000}{60 \times 60}$$

$$\text{Power} = 2725 \text{ kW}$$

Work Diagrams (force–distance graphs)

Work involves both force and displacement and the way they vary in a particular case can be displayed on a force–distance graph. A constant force moving steadily gives a simple straight line graph. (Fig. 5.8.) Note that the work done (force × distance) is equal to the area of the rectangle under the graph.

$$\text{Work done after 20 m} = 40 \times 20 = 800 \text{ J}$$

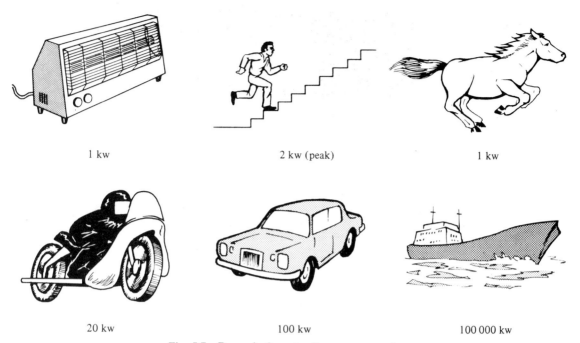

1 kw   2 kw (peak)   1 kw

20 kw   100 kw   100 000 kw

Fig. 5.7 Power is the rate of energy conversion

A constant force moving in a straight line is simple but it is also unusual. Real machines apply varying forces. A simple case is the extension of a spring where the force increases uniformly with distance. Weights added to the spring produce the extensions shown in Fig. 5.9. The average force applied during the extension is $\frac{50}{2}$ newton and the distance moved is 0.4 metre.

$$\text{Work done} = \text{average force} \times \text{distance}$$
$$= \frac{50}{2} \times 0.4 = 10 \text{ J}$$

Note that this is again the area under the graph shaded in Fig. 5.9.

The force distance diagram for a single cylinder engine shows the irregular thrust it gives to the load despite a heavy flywheel which smooths the motion. Alternatively more cylinders will give a more even thrust. (Fig. 5.10.) The Wankel rotary engine is particularly smooth since it does not have pistons which change direction.

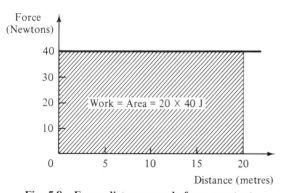

Fig. 5.8 Force–distance graph–force constant

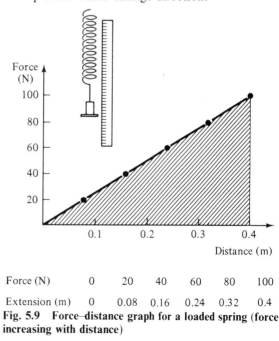

Force (N)	0	20	40	60	80	100
Extension (m)	0	0.08	0.16	0.24	0.32	0.4

Fig. 5.9 Force–distance graph for a loaded spring (force increasing with distance)

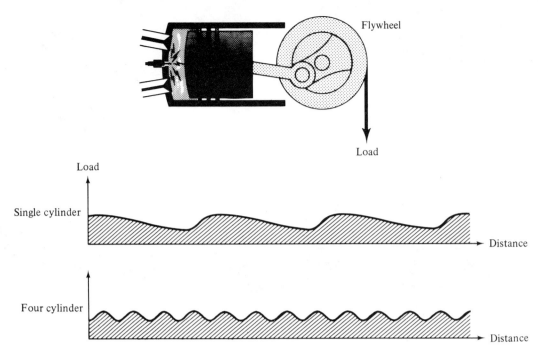

Fig. 5.10 Force distance diagram for a piston-driven hoist

Torque–Angle Graphs

The performance of a rotating engine is best expressed in terms of its turning effect or torque. Torque is calculated in a similar way to moment about the centre of rotation. (Fig. 5.11.)

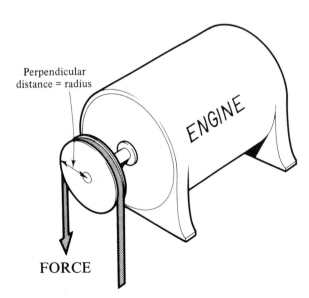

Fig. 5.11 Torque = force × radius of drive pulley

Torque = force × perpendicular distance

The work done by a rotating engine may be calculated from the torque it applies and the angle through which it turns

Work done = torque × angle (Joules)

The angle of rotation must be expressed in the radian measure described on page 19.

Problem
An engine exerts an average torque of 200 Nm and rotates at a speed of 1000 rpm. Calculate the work done in one minute.

Solution
1 revolution = 2π radians
1000 rpm = $1000 \times 2\pi$ radians per minute
Work done in 1 minute
$= 1000 \times 2\pi \times 200 = 1.26 \times 10^6$ J

The torque exerted by an engine may be plotted on a graph against the angle of rotation. (Fig. 5.12.) The torque would rarely be constant with angle and is more likely to vary periodically in some way. (Fig. 5.13.) Note that the area under the curve represents the work done in both graphs.

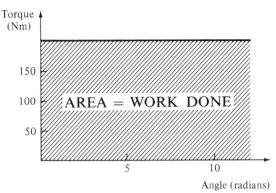

Fig. 5.12 **Torque–angle graph for constant torque**

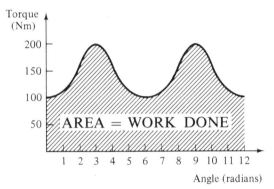

Fig. 5.13 **Torque–angle graph for variable torque**

The Earth's Energy Sources

Although energy may take many forms there are only a few sources and the amount available on earth is limited. The sun is the major source of energy. It keeps the earth warm and supplies the energy for the growth of plants. Animals obtain their energy by eating or burning the plants. The sun also drives the winds and waves and so these provide sources of energy which are renewed daily. Wind, wave or hydroelectric generators could provide almost inexhaustible sources of energy.

The fossil fuels coal, oil, and gas possess chemical energy from the effect of the sun on organic matter accumulated over millions of years. They are not being restored as fast as we are using them up. So far we have discovered new deposits quicker than the growth of demand but it cannot last and we are approaching a crisis.

The radioactive materials in the earth would be used up on a much longer time scale. They were formed at some unknown stage in the development of the earth. Radioactive pollution may limit the extent to which we use nuclear energy. On the other hand if we solve the problems of changing hydrogen into helium under control which requires a temperature of millions of degrees our energy problem will be over. This is the hydrogen bomb process and bearing in mind that all substances are gaseous at these temperatures it is no mean task.

The tides in the oceans derive their energy from the rotation of the earth and moon. Generators driven by the tide could provide energy near coasts. The low water speeds and the stop/start nature of tides make them an inconvenient source of energy. Anyway they would ultimately slow down the rotation of the earth and extend the day!

The deep interior of the earth is hot. Since its formation the crust of the earth has cooled and solidified and it insulates the interior from more rapid cooling. Also the radioactivity of some rocks helps to maintain the interior temperature. If we could pump water to a significant depth it would emerge as steam and provide energy. Even this would cool the earth eventually.

The key to all these possible sources of energy is the time scale. Before each possible source of energy is depleted or produces undesirable effects we must have mastered the technology of harnessing an alternative source. At the present time the situation is critical. The shortage of fossil fuels, particularly oil, is causing economic and political problems. The race is on and during your lifetime fortunes will be won and lost and nations will fade or prosper according to how well energy problems are solved.

Questions

1. What advantages of petroleum fuels cause us to use it extensively despite limited world supplies?

2. If and when oil supplies run out what fuel supplies would you expect to replace them (a) on the medium term, (b) on the long term?

3. Hydroelectric power can be derived from the sea's tides. Where does the energy to drive the tides come from and what would be the long term effect of restricting the tides on a large scale?

4. What powers the waves and the wind?

5. If some new form of energy were discovered, how would we verify that it possessed energy?

Multiple Choice

1. Which has different units from the others?
(a) work, (b) energy, (c) power, (d) heat.

2. Which of the following is *not* a form of energy?
(a) work, (b) torque, (c) electricity, (d) heat.

3. Which is a unit of energy?
(a) newton, (b) joule, (c) kelvin, (d) °C.

4. Work done is equal to . . .?
(a) $\dfrac{\text{force}}{\text{distance}}$, (b) velocity^2,

(c) mass × acceleration, (d) force × distance.

5. Which is the unit of power?
(a) joule, (b) newton, (c) watt, (d) pascal.

6. Which expresses joules in terms of newtons N metres m and seconds s?
(a) m/s, (b) m s, (c) N/m, (d) N m.

7. Which is not a unit of energy?
(a) kilojoule, (b) joule, (c) watt, (d) kilowatt-hour.

8. Power is
(a) work done, (b) rate of working,
(c) energy consumed, (d) maximum force.

9. If the effective power is doubled, the work done during 1 minute will
(a) halve, (b) remain constant, (c) double,
(d) increase fourfold.

10. A particular car engine has a power of 10 kW. When working at full power, the amount of energy it converts in each minute is
(a) 170 J, (b) 10 kJ, (c) 600 kJ, (d) 1 MJ.

Problems

1. A force of 80 N moves through a distance of 15 m. How many joules of work are done? [1200 J]

2. How much energy is expended when a force of 250 N moves through a kilometre? [250 kJ]

3. A girl of weight 500 N in jumping raises her centre of mass by 1.6 m. What is the minimum amount of work she needs to do? If the force was exerted over a time interval of 0.10 s, calculate the average power developed. [800 J; 8 kW]

4. A small object is moved 5 m up the sloping surface of a ramp which measures 4 m horizontally and 3 m vertically. If the object weighs 5 N calculate the work done, neglecting all frictional forces. Calculate the least force parallel to the ramp which could produce this movement. [15 J; 3 N]

5. An engine moves a load of 500 N through 20 m in 5 s. Calculate the work done and the rate of working (power) of the engine. [10 kJ; 2 kW]

6. How many joules of work are expended by a 2 kW motor during 1 hour? [7.2 MJ]

7. Calculate the kinetic energy of a car of mass 1200 kg travelling at 25 m/s. If it brakes to rest in 40 m calculate the braking force. [375 kJ; 9375 N]

8. A motor bike of mass 250 kg is accelerated from rest by a force of 500 N acting over a distance of 12 m. It is then stopped by a braking force of 200 N. Over what distance does the braking force act? Calculate the energy (kinetic) imparted to the bike and hence deduce the maximum speed it reaches. [30 m; 6 kJ; 6.9 m/s]

9. Name five forms of energy and in each case describe how a transition to some other form could occur. If a car of mass 1460 kg travelling at 27.0 m/s is brought to rest by the brakes, how much heat energy is generated? [532 kJ]

10. A spring is loaded by adding five separate masses each of 2.0 kg. Each mass produces an extension of 22 mm. Plot a graph of force in newtons against extension in millimetres. Determine the area under the graph and hence the work done on the spring. Take $g = 9.8$ m/s^2 [5.4 J]

11. A spanner of mass 0.7 kg is dropped from a balloon flying at 3000 m. How much mechanical energy is converted to heat energy before it comes to rest on the ground. Take $g = 9.8$ m/s^2. [21 kJ]

12. The engine of a motor boat produces a useful power of 5.0 kW and is able to drive the boat through still water at a steady speed of 2.0 m/s. Calculate the force opposing the motion of the boat through the water. [2.5 kN]

Multiple Choice Answers

1(c), **2**(b), **3**(b), **4**(d), **5**(c), **6**(d), **7**(c), **8**(b), **9**(c), **10**(c).

6.
Temperature

Everyone has an intuitive idea of the meaning of temperature but we have difficulty when it comes to putting a concise explanation into words. We may describe temperature as the degree of hotness but this is not a definition. Heat and temperature should not be confused. Heat is energy and when energy is added to an object temperature change is just one effect that may be produced. Heat is related to temperature but it is not the same.

One thing we do know is that temperature is the quantity which determines the direction of heat flow. Heat flows from objects at a high temperature to objects at a lower temperature, and where there is no heat transfer we know that the objects are at the same temperature.

Recognizing the difference of temperature is not enough, however; we need to devise a means of defining and measuring temperature precisely. The human body is sensitive to temperature differences, but it is quite unreliable as a means of temperature measurement. The skin is sensitive not so much to temperature, as to heat flow into the skin (warm sensation) or out of the skin (cold sensation).

Certain fixed temperatures we can define, because under constant conditions a substance always changes from solid to liquid, or liquid to vapour at the same temperature. For example, we can talk about the temperature at which mercury freezes and know that we are referring to a specific temperature. This change of state can be used to classify temperatures as shown in Figs. 6.1 and 6.2. We could specify the temperature at which ice melts or water boils in the same way, but how do we specify temperatures in between these points?

As a substance gets hotter, certain properties

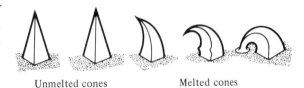

Unmelted cones   Melted cones

Fig. 6.1   Clay cones having different melting points are used in ovens to indicate that the temperature lies between certain limits

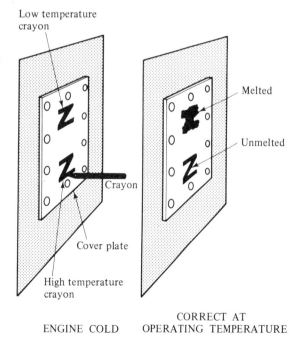

Low temperature crayon

Melted

Unmelted

Crayon

Cover plate

High temperature crayon

ENGINE COLD

CORRECT AT OPERATING TEMPERATURE

Fig. 6.2   Use of temperature sensitive crayons to check temperatures

change. Most substances, when they are heated, expand in size, change their electrical resistance, and, if they are glowing hot, change colour. We can use the variation in a property of a material to bridge the gap between any fixed points. In fact, we find that, using a varying property in this way, we need only two fixed points to define any temperature. The two fixed points we choose are the melting point of ice and the boiling point of water. The property which is most commonly used to measure temperature is the expansion of mercury in a glass tube.

Mercury-in-Glass Thermometer

This thermometer consists of a sealed glass tube of narrow bore with a bulb at one end containing mercury, which fills the bulb and extends up the tube. To calibrate the thermometer, we immerse the bulb first in melting ice, and then in the steam from water boiling at standard atmospheric pressure, in each case marking the level of the mercury on the stem. (Fig. 6.3.) The distance between the two fixed points is called the fundamental interval of the thermometer, and it is divided up into 100 equal parts or degrees.

The scale of temperature with zero at the freezing point (0°C) and 100 degrees at the boiling point (100°C) is called the Celcius scale after the Swedish physicist who devised the mercury thermometer 250 years ago.

Accuracy of Mercury Thermometer

Mercury-in-glass thermometers are subject to errors which are difficult to eliminate. For example, the glass bulb and stem, being slightly plastic, take some time to contract when cooled. The pressure of the mercury thread causes different readings in the vertical and horizontal positions, and irregularities in the bore produce variations between thermometers in the size of degrees. Mercury thermometers are very convenient, however, and we can achieve an accuracy of 0.01°C by using a thermometer which has been calibrated against an accurate thermometer at the National Physical Laboratory.

The range of the mercury thermometer extends from the freezing point to the boiling point of mercury, −39°C to 356°C. Alcohol-in-glass thermometers are used to measure atmospheric temperatures below −39°C. More accurate and more convenient thermometers than the liquid-in-glass variety exist and are preferred for many applications in science and industry.

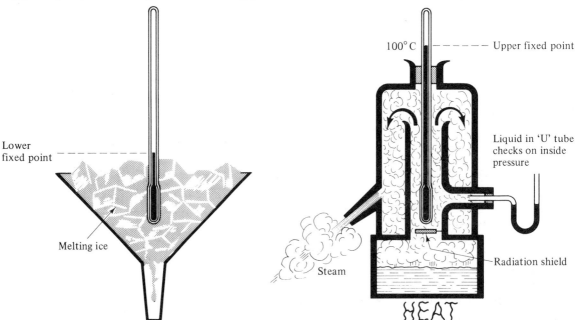

Fig. 6.3 **Determination of the fixed points of a mercury-in-glass thermometer**

Absolute Zero and the Kelvin Scale

The freezing point of water is quite an arbitrary point at which to fix our zero of temperature. Is there a more natural zero we could use? The answer is yes. There does exist an absolute zero of temperature at which no more internal energy can be extracted from a body.

This absolute zero of temperature occurs at $-273°C$. It makes sense to have a scale of temperature in which the lowest temperature obtainable has a value of zero. This scale, called the Kelvin scale, uses the same unit as the Celsius scale but just starts from a different point. (Fig. 6.4.)

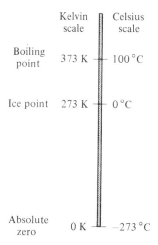

Fig. 6.4 **The Kelvin and Celsius scales start from different points**

Note that the term degree is dropped when specifying Kelvin temperatures, e.g., 100 degrees Celsius ($100°C$) becomes 273 Kelvin (273 K). Differences in temperature whether they be on the Celsius or Kelvin scale are expressed in Kelvin, e.g.,

$$50°C - 20°C = 30 K$$

Problem

Express $150°C$ on the Kelvin scale.

Solution

$$\text{Kelvin temp} = °C + 273$$
$$= 150 + 273 = 423 K$$

$150°$ is the same temperature as 423 K.

The Platinum Resistance Thermometer

A rise in the temperature of a metal produces an increase in its electrical resistance. Since we can measure resistance very accurately, we can use this property as an accurate measure of temperature. Pure platinum is commonly used as the resistive material, being extremely unreactive chemically. The resistance of platinum increases by about 40 per cent between $0°C$ and $100°C$.

The platinum resistance thermometer consists essentially of a fine platinum wire coil wound on a mica former. (Fig. 6.5.) A sensitive ohmmeter serves to measure platinum resistance as it varies with temperature.

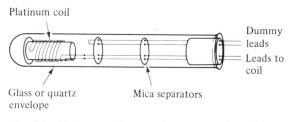

Fig. 6.5 **Platinum resistance thermometer. To eliminate fringe effects the difference in resistance of the dummy and main leads is measured**

The range of the platinum resistance thermometer extends from $-200°C$ up to $1500°C$ (60 K to 1800 K) if high-melting point materials are used in its construction. It is highly suitable for measuring the high temperatures of heat-treating and annealing processes in the metallurgical industry, and it adapts easily for automatic recording or control systems. It has the disadvantage of responding slowly to temperature variation. This is one of the reasons why it is not used so widely in industry as the thermo-electric thermometer.

Thermocouple

At the point of contact between two different metals there exists an electrical potential difference which depends on the temperature of the junction. When we complete the circuit with a second junction at a different temperature, a current flows round the circuit. (Fig. 6.6.) This thermoelectric effect is called the Seebeck effect after the man who

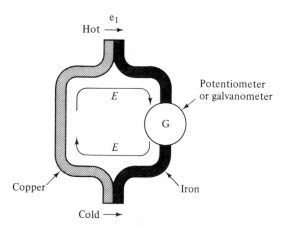

Fig. 6.6 Thermoelectric e.m.fs. are generated at the junction between different metals

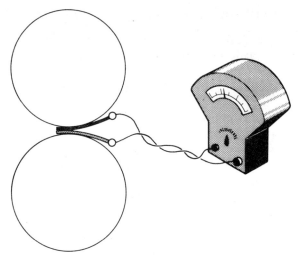

Fig. 6.7 Thermocouples can measure temperature in places inaccessible to other thermometers, such as the nip of heated rollers

discovered it, and the junction is called a thermocouple. The resulting current may be used to measure temperature. For less accurate measurements, we can dispense with the cold junction completely, and replace it by the measuring instrument. (Fig. 6.7.) The thermocouple then behaves as though its cold junction were at room temperature.

The robustness, low thermal capacity, and compact size of the thermocouple make it the most widely used thermometer in industry. It lends itself well to remote control and to automated systems. By choosing suitable combinations of metals or alloys, such as platinum-rhodium alloy, we can measure temperatures up to 1500°C. Other common combinations (thermels) are iron-constantan and chromel-constantan. The sensi-

tivity which can be achieved by combining thermocouples into a thermopile Fig. 6.8 must be seen to be believed. Such an instrument with several hundred junctions can detect the heat of a candle flame thirty metres away.

Temperature Coefficients

Several physical properties, such as length, pressure and electrical resistance, vary with temperature, and we must make allowance for the variation in the design of a countless number of devices. (Fig. 6.9.)

The change in a property is not always directly

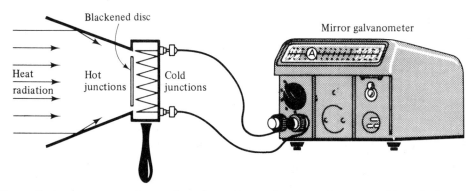

Fig. 6.8 In a radiation pyrometer, a thermopile made up of many thermocouples connected in series is used to measure the heat radiation

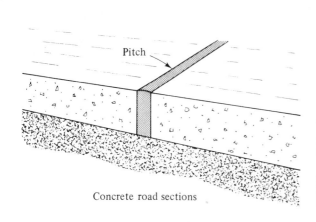

Concrete road sections

Power cable sag

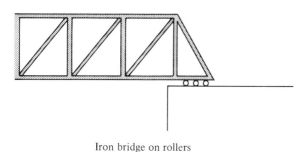

Iron bridge on rollers

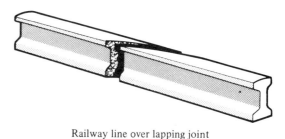

Railway line over lapping joint

Fig. 6.9   Allowances in structures for thermal expansion

proportional to the change in temperature, but we can describe all those changes which are approximately so, in a similar way. The variation in each property is described by a temperature coefficient, defined as the fractional change in the property per unit rise in temperature.

We will illustrate the method of approach for the specific case of linear expansion. (Fig. 6.10.)

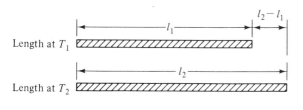

Difference in temperature $= (T_2 - T_1) = \theta$

Difference in length $= (l_2 - l_1)$

6.10   Linear expansion (exaggerated)

The coefficient of linear expansion α is the fractional change in length per unit rise in temperature.

$$\alpha = \frac{\text{change in length}}{\text{original length} \times \text{temperature rise}}$$

$$\alpha = \frac{l_2 - l_1}{l_1 \times \theta}$$

Change in length
$$= l_1 \alpha \theta$$
New length = original length + change in length
$$l_2 = l_1 + l_1 \alpha \theta$$
or
$$l_2 = l_1(1 + \alpha \theta)$$

One of the many forms of apparatus used to measure α is shown in Fig. 6.11. The length at room temperature is measured by an accurate steel rule and the change in length is the difference between two depth gauge readings.

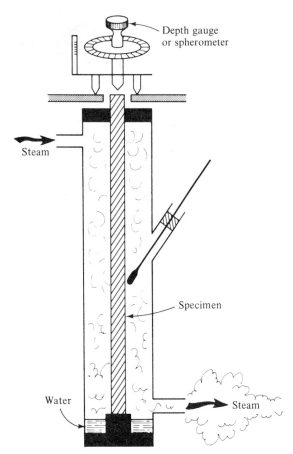

Depth gauge
or spherometer

Steam

Specimen

Water

Steam

Fig. 6.11 Measurement of the linear expansion coefficient over the range from room temperature to 100°C

Problem
An iron girder of length 5 metres is subject to temperatures of $-5°$ to $+35°$. What is the change in length? $\alpha = 12 \times 10^{-6}$ K

Solution
$$\text{Change in length} = l_1 \alpha \theta$$
$$= 5 \times 12 \times 10^{-6} \times 40$$
$$= 24 \times 10^{-3} \text{ m} = 2.4 \text{ mm}$$
$$\text{Change in length} = 2.4 \text{ mm}$$

Volume Coefficient γ

The thermal expansion of liquids is described by a volume coefficient, γ, i.e., the fractional change in volume per degree rise in temperature.

$$\text{Change in volume} = \text{original volume} \times \gamma$$
$$\times \text{ temperature change}$$
$$= v_1 \gamma \theta$$

There is a simple relationship between linear and volume coefficients of the same material:

$$\gamma = 3\alpha$$

This arises because the volume extends in three dimensions. Note that values of α are quoted for solids, and values of γ for liquids and gases in Table 6.1.

Table 6.1 Coefficients of Expansion in Range 0–100°C

Solids	α (/K)	
Aluminium	23.8	$\times 10^{-6}$
Brass	18.7	$\times 10^{-6}$
Copper	16.8	$\times 10^{-6}$
Glass (soft)	8.5	$\times 10^{-6}$
Iron	12	$\times 10^{-6}$
Plaster	10	$\times 10^{-6}$
Platinum	9	$\times 10^{-6}$
Wood along grain	$\simeq 8$	$\times 10^{-6}$
across grain	$\simeq 40$	$\times 10^{-6}$
Liquids	γ	
Glycerine	0.51	$\times 10^{-3}$
Mercury	0.18	$\times 10^{-3}$
Petrol	0.95	$\times 10^{-3}$
Water	0.207	$\times 10^{-3}$
Gases		
Air	3.67	$\times 10^{-3}$
Carbon dioxide	3.72	$\times 10^{-3}$
Hydrogen	3.66	$\times 10^{-3}$
Nitrogen	3.67	$\times 10^{-3}$
Water vapour	4.19	$\times 10^{-3}$

There is a slight complication when dealing with the expansion of liquids in containers. The container also expands and makes the expansion of the liquid appear to be less than it really is. What we observe is the difference between the expansion of the liquid and the expansion of the container. This is easily allowed for using the relation:

$$\gamma_{\text{absolute}} = \gamma_{\text{apparent}} + \gamma_{\text{container}}$$

Problem
A volume of 500 cm³ of mercury is contained in a glass tube graduated in cubic centimetres. What increase in volume will be registered when the

temperature is increased from 20°C to 200°C, if the average coefficients of expansion over this temperature range are

$$\gamma = 0.182 \times 10^{-3}/°C \text{ for mercury}$$
$$\alpha = 8.3 \times 10^{-6}/° \text{ for glass}$$

Solution

$$\gamma_{glass} = 3\alpha = 3 \times 8.3 \times 10^{-6}/°C$$
$$= 24.9 \times 10^{-6}/°C$$

Apparent coefficient

$$= \gamma_{apparent} = \gamma_{mercury} - \gamma_{glass}$$
$$= 0.182 \times 10^{-3} - 24.9 \times 10^{-6}$$
$$= 0.157 \times 10^{-3}$$

Apparent expansion

$$= v_1 \gamma_{app} \theta$$
$$= 500 \times 0.157 \times 10^{-3} \times 180 = 14.1 \text{ cm}^3$$

Temperature Coefficient of Resistance σ

The temperature coefficient of resistance is defined as the fractional increase in resistance per unit rise in temperature. The resistance of metals and alloys increases with temperature while that of insulators decreases. Therefore σ is positive for metals and negative for insulators. (Table 6.2.)

$$\sigma = \frac{\text{increase in resistance}}{\text{initial resistance} \times \text{temperature rise}}$$

$$\text{Change in resistance} = R_1 \sigma \theta$$
$$\text{New resistance} = R_1(1 + \sigma\theta)$$

Table 6.2   Temperature Coefficients of Resistance at 20°C

Material	σ (/K)
Aluminium	3.8×10^{-3}
Brass	2.0×10^{-3}
Carbon (graphite)	-0.5×10^{-3}
Copper	3.9×10^{-3}
Eureka	0.05×10^{-3}
Nichrome	0.4×10^{-3}
Platinum	0.3×10^{-3}

Problem

A copper coil has a resistance of 6.2 ohms at 20°C. What will be its resistance at its working temperature of 80°C? (α for copper $= 3.9 \times 10^{-3}$/K.)

Solution

New resistance = original resistance + change
$$= R_1 + R_1 \alpha \theta$$
$$= 6.2 + 6.2 \times 3.9 \times 10^{-3} \times 60$$
$$= 6.2 + 1.45$$
$$= 7.65 \text{ ohms}$$

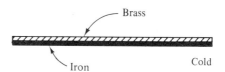

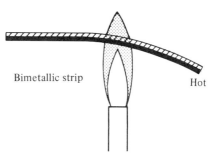

When heated the brass expands more than the iron and the strip becomes curved

(a) BIMETALLIC STRIP

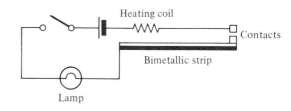

(b) DIRECTION INDICATOR FLASHER UNIT

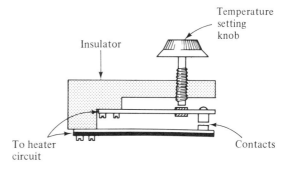

(c) THERMOSTAT

Fig. 6.12   Applications of the bimetallic strip

RIVETTING

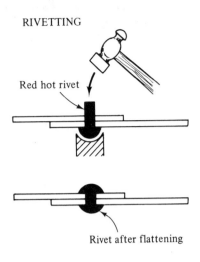

Red hot rivet

Rivet after flattening

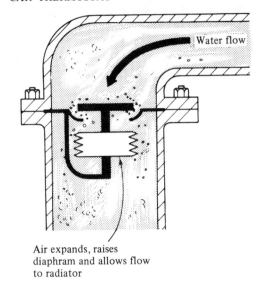

Water flow

Air expands, raises
diaphram and allows flow
to radiator

Fig. 6.13 Application of thermal expansion

Use of Thermal Expansion

In many cases of small scale elastic structures, the effect of thermal expansion is small and it can be ignored. The errors in a metre rule are imperceptible to a carpenter but the expansion in a 30 m surveyor's tape becomes significant. Thus in every case of measurement or design, consideration must be given to thermal expansion. Measuring instruments are usually calibrated at 20°C and where we cannot conduct a measurement at that controlled temperature, we make a calculated compensation.

Where steelwork and masonry form part of a structure as in a bridge or tower block, clearance must be left at the ends of metal girders to allow for the greater expansion of the metal. Bridges are set on rollers to permit movement without fracture. Gaps are left in railway lines and concrete roads and the sag of overhead cables is carefully worked out. (Fig. 6.9.) Without these allowances, the force set up by constrained expansion is enormous and very soon exceeds the strength of the system resulting in permanent damage. Material which is not rigidly secured may move bodily after repeated hot–cold cycles. The movement, called creep, is what happens to lead sheeting on sloping roofs. On expanding the bottom is pushed down,

and on contraction the top is pulled down each time by a minute amount.

In smaller systems such as clocks and timing mechanisms special alloys having extremely low expansion coefficients may be used. However thermal expansion is not always a design problem, it can be used to advantage in a number of ways some of which are shown in Figs. 6.12 to 6.13.

Questions

1. Why does there seem to be a lower limit for temperatures but no upper limit?

2. What are the advantages and disadvantages of a mercury-in-glass thermometer as an instrument for measuring temperature?

3. What happens to the volume of the space inside a steel tank when its temperature is raised?

4. Suggest five examples where the engineer should take account of the effect of thermal expansion.

5. What are the advantages and disadvantages of the human body as a temperature sensor?

Multiple Choice

1. The upper fixed point of a mercury thermometer occurs at
    (a) 0°C, (b) 100 K, (c) 100°C, (d) 273 K.

2. The temperature of a hot summer's day is approximately
(a) 30 K, (b) 80 K, (c) 300 K, (d) 400 K.

3. Which thermometer could *not* be used at $-100°C$?
(a) mercury, (b) alcohol, (c) thermocouple,
(d) resistance thermometer.

4. Which of the thermometers in 3 could *not* be used to measure 150°C?

5. A mercury-in-glass thermometer will be more sensitive to a small change in temperature if it has a
(a) thin-walled bulb, (b) large bore,
(c) long stem, (d) large bulb.

6. The temperatures of the melting point of ice and the boiling point of water respectively are
(a) 0 K and 100 K, (b) 0 K and 273 K,
(c) 100 K and 373 K, (d) 273 K and 373 K.

7. Alcohol is often preferred to mercury as a thermometric liquid because it has a
(a) higher freezing point, (b) higher boiling point,
(c) lower freezing point, (d) lower boiling point.

8. Which type of thermometer is the most suitable for measuring the temperature of a sample of mercury of volume 1.0×10^{-6} m^3?
(a) electrical resistance, (b) thermocouple,
(c) mercury-in-glass, (d) alcohol in glass.

9. A thick glass tumbler is likely to crack if boiling water is poured into it because
(a) thick glass is weak, (b) glass is brittle,
(c) the inside expands more than the outside,
(d) glass is a good conductor of heat.

10. When the temperature of a metal is increased what property is reduced?
(a) volume, (b) electrical resistance,
(c) density, (d) mass.

Problems

1. Express the following temperatures on the Kelvin scale: 0°C, 30°C, $-30°C$, 250°C. [273 K; 303 K; 243 K; 523 K]

2. Express the following temperatures on the Celsius scale: 0 K, 200 K, 300 K, 500 K. [$-273°C$; $-73°C$; 27°C; 227°C]

3. What temperature has the same numerical value on the Celsius and Kelvin scales? [136.5 K]

4. A metal rod 4.0 m long is heated from 10°C to 30°C. Calculate the increase in its length. The coefficient of linear expansion of the metal is 0.000 018/K. [1.4 mm]

5. A steel wire of length 300 m is cooled to 0° from 60°. What is its change in length? (Coefficient of linear expansion of steel $\alpha = 1.2 \times 10^{-5}$/K.) [0.22 m]

6. A steel shaft has a diameter of 50.00 mm at 10°C. What diameter would it have at 50°C? (α steel = 1.2×10^{-5}/K.) [50.02 mm]

7. A metal bar measures 0.6240 m at 20° and 0.6251 m at 100°C. Calculate its coefficient of linear expansion. [2.2×10^{-5}/K]

8. What is meant by the apparent coefficient of volume expansion of a liquid? Calculate its value for mercury-in-glass if the absolute coefficient of volume expansion of mercury = 1.80×10^{-4}/K and coefficient of linear expansion of glass = 8.5×10^{-6}/K. [1.545×10^{-4}/K]

9. A litre of mercury exactly fills a stoppered glass vessel of 1.000 litre capacity at 25°C. What will be the volume of the space above the mercury at 0°C? Use data from Question 8. [3.86×10^{-3} litres]

10. Calculate the increase in resistance of a copper coil of 2.2 ohm for a rise in temperature of 80 K. (Coefficient of resistance for copper $\sigma = 3.9 \times 10^{-3}$/K.) [0.686 ohms]

11. The resistance of a platinum coil is 13.50 ohm in melting ice and 14.70 ohm in steam at normal atmospheric pressure. What is its temperature when its resistance is 13.69 ohm? [16°C]

12. The following table gives the e.m.f.'s in a copper–iron thermocouple circuit for various temperatures of the hot junction (the cold junction is at 0°C).

Temp. (°C)	e.m.f. (mV)	Temp. (°C)	e.m.f. (mV)
0	0	300	2.05
50	0.62	350	1.87
100	1.15	400	1.60
150	1.60	450	1.15
200	1.90	500	0.60
250	2.10		

Plot a graph and deduce the two temperatures giving an e.m.f of +1.75 mV. How would you tell which was the correct temperature in practice? [Approximately 170°C, 375°C]

Multiple Choice Answers

1(c), 2(c), 3(a), 4(b), 5(ad), 6(d), 7(c), 8(b), 9(bc), 10(c).

7.
Heat Energy

A man expends about 12 million joules of energy a day which he could produce by eating about one kilogram of sugar. The same energy could be supplied by a 1 kilowatt electric fire in four hours or by burning half a litre of petrol or half a cubic metre of North Sea gas. (Fig. 7.1.) They are all source of the same thing, energy. No matter what form it takes energy is measured in terms of the same unit; the joule.

The realization that all forms of energy are interchangeable did not happen overnight. Like many revolutionary discoveries it had to be fought for with experimental evidence and argument.

The Caloric Theory of Heat

Two hundred years ago the prevailing theory among scientists was that heat was an invisible fluid. They thought that this fluid, which they called caloric, filled up the spaces between the molecules of a body and flowed from hot bodies into cold bodies when they came into contact. They attributed the fact that bodies expand when heated to the increased quantity of caloric that flowed into them. The caloric theory worked reasonably well in accounting for heat flow. Early engineers visualized heat engines as operating on a flow of caloric, in the same way as a hydraulic turbine operates on a flow of water.

Count Rumford, an international adventurer and engineer, who was engaged in boring out cannon barrels for the Bavarian Army dealt the caloric theory a severe blow. He attacked the assumption that the rise in temperature produced during sawing or drilling was a result of the caloric squeezed out of the sawdust or grindings. By using a blunt drill, which removed very little metal, he showed that the heat had more to do with energy expended by the horse which turned the drill than with the drillings produced.

Molecular Vibration Theory of Heat

The modern theory of the nature of heat associates heat with mechanical energy. The internal energy

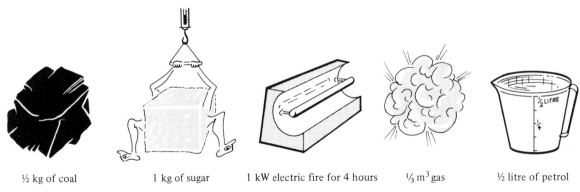

| ½ kg of coal | 1 kg of sugar | 1 kW electric fire for 4 hours | ⅓ m³ gas | ½ litre of petrol |

Fig. 7.1 Approximate equivalents to the energy expended by a man in one day

of a body is the sum total of the kinetic and potential energy of vibration of the molecules making up the body. (Fig. 7.2.) When heat is added to a body its molecules vibrate more energetically. It is this increased vibration that causes the effects, such as thermal expansion, that we associate with temperature rise. The vibrations can be increased by contact with a body at a higher temperature or by mechanical activity such as hammering or rubbing.

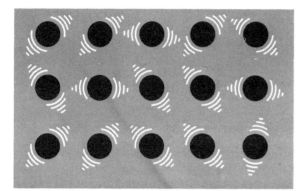

Fig. 7.2 Molecules of a solid vibrating with thermal energy

Once heat became accepted as a form of energy, the principle of the conservation of energy could be formulated. It states **that energy cannot be created or destroyed but only converted from one form to another.** Thus, heat can be converted into or derived from mechanical, electrical, or kinetic energy, etc.

Calorific Values

To choose the best fuel for a particular purpose, we must consider very carefully such factors as the cost, the convenience, and the availability of the fuel. In assessing the cost, it is important to know the **calorific value** of the fuel, i.e., **the heat of combustion per unit mass.** These are quoted in Table 7.1 for some common fuels and foods.

We measure calorific values by burning a measured quantity of fuel in a strong steel container, called a bomb calorimeter. (Fig. 7.3.) An electrical filament ignites the fuel, and, to ensure complete combustion, the calorimeter is filled with oxygen under pressure. The heat produced increases the temperature of the bomb calorimeter

Table 7.1 Calorific Values of Fuels and Foods
Round figures are quoted because large variations can occur in the quality of the sample

	Calorific value (J/kg)
Fuels	
Coal	30×10^6
Fuel oil	46×10^6
Manufactured gas	28×10^6
Natural gas	55×10^6
Petrol	46×10^6
Wood	14×10^6
Foods	
Bread	11×10^6
Butter	34×10^6
Milk	3.0×10^6
Potatoes	40×10^6
Sugar	16×10^6

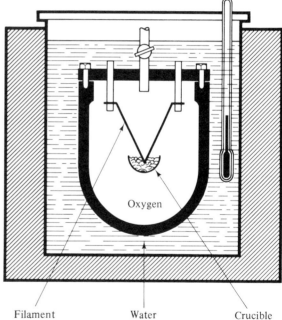

Filament   Water   Crucible

Fig. 7.3 In the bomb calorimeter fuel is burnt in oxygen under pressure and the heat produced is measured

and the surrounding water, all of which is in an insulated container to prevent loss of heat. Having made allowances for the heat supplied by the filament, and the heat absorbed by the calorimeter, we can calculate the calorific value of the fuel.

65

Specific Heat Capacity C

When we heat equal masses of copper and aluminium in the same flame, the temperature of the copper rises about twice as fast as that of the aluminium. The aluminium accepts the same amount of heat but shows less of a change in temperature. We say that the aluminium specimen has a greater **heat capacity** than the copper specimen. As usual, we want to express this property of a substance in a precise way, basing it on a unit mass of the substance and unit temperature change. **This property is called specific heat capacity, and it is defined as the heat required to raise 1 kilogram of a substance by 1 Kelvin.** (Note that whenever we use the word 'specific' before a quantity we mean 'per unit mass'.) The symbol for specific heat capacity is C and its value for some common materials is given in Table 7.2.

Table 7.2 **Specific Heat Capacities (J/kg K)**

Substance	Sp. ht. caps.	Substance	Sp. ht. caps.
Air*	7.1×10^3	Iron	0.5×10^3
Ethyl alcohol	2.5×10^3	Lead	0.13×10^3
Aluminium	0.92×10^3	Mercury	0.14×10^3
Copper	0.39×10^3	Steam*	1.4×10^3
Glass	0.13×10^3	Marble	0.88×10^3
Hydrogen*	10×10^3	Water	4.2×10^3
Ice	2.1×10^3	Wood	17×10^3

* The values quoted are for the case of constant volume.

From the definition of specific heat:

Specific heat capacity C

$$= \frac{\text{quantity of heat energy } (Q)}{\text{mass of substance } (m) \times \text{temperature change } (\theta)}$$

$$C = Q/m\theta$$

We use this equation to calculate quantity of heat in the form

$$Q = mC\theta$$

Q is given in joules when m is in kilograms, θ is in Kelvin, and C is in J/kg K.

Problem

An aluminium pan of mass 0.5 kg and containing 3 kg of water is heated from 10°C to 100°C. How much heat has been received by the pan and contents? C for aluminium $= 0.92 \times 10^3$ J/kg K.

Solution

Using equation $Q = mC\theta$, quantity of heat received by pan, Q_p is

$$Q_p = 0.5 \times 0.92 \times 10^3 \times (100 - 10) \text{ J}$$
$$= 4.1 \times 10^4 \text{ J}$$

Quantity of heat received by water, Q_w is

$$Q_w = 3 \times 4.2 \times 10^3 \times (100 - 10) \text{ J}$$
$$= 113 \times 10^4 \text{ J}$$

Total heat received $= Q_p + Q_w = 117 \times 10^4 \text{ J}$
Total heat received $= 1.17 \text{ kJ}$.

Measurement of Specific Heat

The simplest way of measuring the specific heat of a solid is to heat it and then immerse it in a calorimeter containing say W kg of water. Heat flows from the specimen into the water and calorimeter until they are all at the same temperature.

Using the symbols of Fig. 7.4, heat gained by calorimeter and contents = heat lost by specimen, i.e.,

$$m_c C_c (T_2 - T_1) + W C_{\text{water}} (T_2 - T_1) = mC(100 - T_2)$$

We calculate the specific heat capacity of the specimen from this equation. When the heat capacity of the calorimeter is small compared to that of the water it contains, we neglect the term $m_c C_c (T_2 - T_1)$.

Measuring the Specific Heat of Liquids

We can use the method of mixtures to measure the specific heats of liquids by immersing a hot solid of known specific heat in a quantity of the liquid. Equating the heat lost and the heat gained we have

$$m_c C_c (T_2 - T_1) + m_{\text{liquid}} C_{\text{liquid}} (T_2 - T_1) = mC(100 - T_2)$$

In a more convenient method, the liquid is heated by an electrical element. (Fig. 7.5.) The electrical

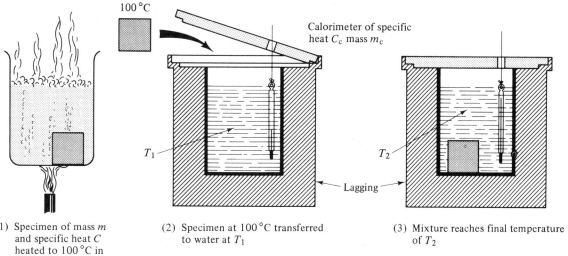

100 °C

Calorimeter of specific
heat C_c mass m_c

T_1

Lagging

T_2

(1) Specimen of mass m
and specific heat C
heated to 100 °C in
boiling water

(2) Specimen at 100 °C transferred
to water at T_1

(3) Mixture reaches final temperature
of T_2

Fig. 7.4   Measurement of specific heat by the method of mixtures

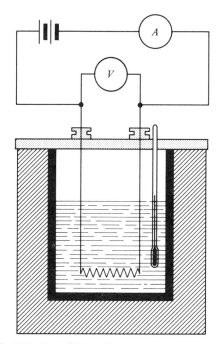

A

V

Fig. 7.5   Liquid heated by an electrical element

energy supplied, VIt joules, is absorbed by the calorimeter and the contained liquid.

$$m_c C_c (T_2 - T_1) + m_{liquid} C_{liquid}(T_2 - T_1) = VIt$$

Alternatively, the heat can be continuously absorbed by a flow of liquid. (Fig. 7.6.) After the liquid has flowed for some time at a constant rate,

the thermometers give steady readings and if a mass m flows in a time t then

$$m_{liquid} C_{liquid}(T_2 - T_1) = VIt$$

The vacuum jacket reduces heat loss to a low value and, since the temperature of the calorimeter remains constant, its heat capacity does not enter into the calculation.

Latent Heat L

When heat is continually supplied to a solid its temperature rises. Once it begins to melt the temperature remains constant until all the solid becomes liquid. (Fig. 7.7.) The continuing supply of heat then causes the temperature to rise until the boiling point is reached. Again the temperature stabilizes until all the liquid has vaporized after which the temperature of the vapour rises further.

The heat which produces the temperature change is referred to as sensible heat because the change is readily apparent on a thermometer. During the changes from solid to liquid and liquid to vapour the heat supplied is absorbed by the substance in re-arranging its internal structure. In fact the molecules are gaining potential energy. The heat used in this way is not indicated by the thermometer and it is referred to as latent (hidden) heat. **The specific latent heat of a substance is the**

67

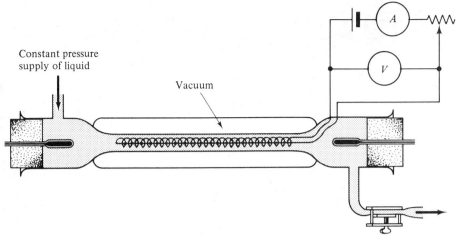

Fig. 7.6 Heat continuously absorbed by a flow of liquid

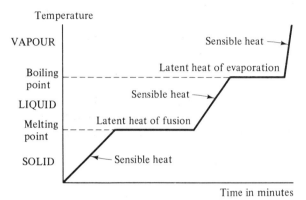

Fig. 7.7 When a substance is supplied with heat at a constant rate, the temperature rise is delayed at the melting and boiling points, due to the absorption of latent heat

Table 7.3

Substance	Specific latent heat of fusion (J/kg)	Specific latent heat of evaporation (J/kg)
Carbon dioxide	190×10^3	367×10^3
Glycol	182×10^3	802×10^3
Mercury	11.8×10^3	286×10^3
Naphthalene	8.4×10^3	297×10^3
Oxygen	13.9×10^3	214×10^3
Water	335×10^3	2270×10^3
Aluminium	400×10^3	—
Copper	200×10^3	—
Gold	67×10^3	—
Silver	105×10^3	—
Iron	210×10^3	—

heat required to convert unit mass of the substance from one state to another at a constant temperature. Each substance has a latent heat of fusion (melting) and a latent heat of evaporation. Specific latent heats are quoted in joules per kilogram per Kelvin at either the melting point or the normal boiling point. The graph of Fig. 7.7 shows quite clearly the pause in temperature rise as the heat is used to change the solid to liquid, and the liquid to the vapour. A substance cooling through its melting or boiling point maintains its temperature until the latent heat is transmitted to its surroundings (one reason why steam and hot pitch give nasty burns).

Problem

How much heat energy is required to melt 5 kg naphthalene at its melting point? (Specific latent heat of naphthalene $L = 8.4 \times 10^3$ J/kg.)

Solution

$$\text{Heat required} = \text{mass} \times \text{Sp latent heat}$$
$$= 5 \times 8.4 \times 10^3$$
$$\text{Heat required} = 4.2 \times 10^4 \text{ J}$$

Problem

Two kilograms of ice cubes at 0°C are heated until they melt. The heating is continued until the water reaches the boiling point and then boils dry. How much heat is required? Take $L = 335 \times 10^3$ J/kg.

68

Solution

Heat to melt $= mL = 2 \times 335 \times 10^3$
$= 6.7 \times 10^5$ J

Heat to raise from $0°$C to $100°$C
$= Q = mc\theta$
$= 2 \times 4.2 \times 10^3 \times 100$
$= 8.4 \times 10^5$ J

Heat to boil to dryness
$= mL = 2 \times 2270 \times 10^3$
$= 45.4 \times 10^5$

Total heat required
$= 6.7 \times 10^5 + 8.4 \times 10^5 + 45.4$
$\times 10^5$

Heat required $= 60.5 \times 10^5$ J

Thermal Properties of Water

Water is unique in being the commonest substance on the surface of the earth and the basic constituent of living matter. It also has some outstanding thermal properties which help to keep the environment within tolerable limits of temperature.

Firstly water has a very high specific heat and so comparatively large amounts of heat must flow in or out of it before its temperature is changed appreciably. It is an excellent medium for storing heat. The heat absorbed by lakes and oceans in the summer is released during the winter and this reduces the extremes of temperature in both seasons. Thus the climate of Edinburgh on an island is very different from that of Moscow well inland although they are situated on the same latitude.

Engineers use the high thermal capacity of water in heat exchangers ranging from boilers to hot water bottles. Although water is corrosive it is cheap, readily available and non-inflammable and hence it is universally used for transferring heat in industry. When a liquid is heated, the increased amplitude of vibration of its molecules increases their average separation and the liquid expands. Water is an exception to this and actually contracts as its temperature rises from $0°$C to $4°$C. The behaviour of water near its freezing point is very important for animals and plants in water. As the water cools at the surface, it becomes denser and falls to the bottom. This produces convection currents which continue until the water reaches a temperature of $4°$C at which temperature water has a maximum density. Any further cooling below $4°$C produces expansion and the cooled water remains on the surface. Eventually, ice forms and, being less dense than the water, it floats on the surface and insulates the underlying water from further heat loss. Thus, the water at the bottom of ponds and lakes may remain at $4°$C all throughout a long cold winter permitting the survival of underwater life. (Fig. 7.8.)

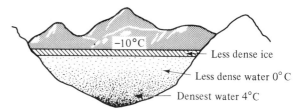

Fig. 7.8   Unless it freezes solid, the water at the bottom of a lake will be at $4°$C

Heat Engines

A heat engine operates by changing heat energy into a mechanical energy, and our three main requirements of a heat engine are:

(a) High efficiency
(b) Large ratio of power to size
(c) Operation on cheap fuel

We will look at some common heat engines in the light of these requirements and ignore, for the moment, other important qualities (such as long life, quietness in operation, low frictional losses, and ease of manufacture), which are common to all machines.

Efficiency of An Engine

Of the heat supplied to an engine only a fraction is converted into useful work. The rest is lost to the surroundings through friction, heat conduction and the exhaust gases. (Fig. 7.9.) The efficiency of the engine is measured by the ratio of the useful work output to the total energy input.

$$\text{Efficiency} = \frac{\text{useful work done}}{\text{energy supplied}} \times 100\%$$

69

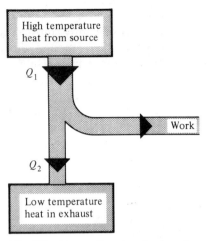

Fig. 7.9 Energy flow in a heat engine

A machine such as a gearbox taking in mechanical energy and transferring it to some other point may achieve a high efficiency like 90%. If the whole process of conversion of an engine is considered however, much lower efficiencies are normal. A car engine would typically deliver only about 15% of the energy of the petrol as useful work at the wheels.

Problem

An engine burns 40 kg of fuel oil of calorific value 46×10^6 J/kg and performs 7.2×10^8 J of work. Calculate its efficiency.

Solution

Energy supplied $= 40 \times 46 \times 10^6 = 18.4 \times 10^8$ J

$$\text{Efficiency} = \frac{\text{work done}}{\text{energy supplied}}$$

$$= \frac{7.2 \times 10^8}{18.4 \times 10^8} \times 100\%$$

Efficiency $= 39\%$

Thermodynamics

The study of the relation of heat energy to other forms of energy is called thermodynamics.

Over 100 years ago, Sadi Carnot, a young French engineer, approached the problem of steam engine design in a theoretical but simple way. He deduced theoretically that the efficiency depended on having a high inlet temperature and a low exhaust temperature. His discovery produced a leap forward in steam engine design and laid the foundations of the study of heat engines. Before we describe the design and performance of particular engines, we will examine some basic principles of thermodynamics relevant to all engines.

The First Law of Thermodynamics

When the principle of conservation of energy is stated with reference to heat and work, it is called the first law of thermodynamics. **The heat supplied to a system is equal to the rise in internal energy plus the work done by the system on its surroundings.**

Heat supplied = increase in internal energy
+ work done

The increase in internal energy means an increase in the kinetic and potential energy of the molecules of a body which shows itself as a change in temperature or a change in state. The work done involves the movement of a force.

The Second Law of Thermodynamics

If we could extract energy from cold places (making them colder) and use the energy to drive engines or heat homes there would never a world shortage of energy. The first law of thermodynamics (i.e., energy is conserved) does not forbid this. However, no engine has ever managed to do it. So we have a second law of thermodynamics which emphasizes the fact. **A machine cannot do useful work by extracting heat from the coldest part of its surroundings.** It predicts that the efficiency of an engine cannot equal 100%, and that heat cannot be made to flow from a hot to a cold body unless extra energy is supplied, e.g., as in a re-frigerator. It means that we must go on providing our engines with expensive high temperature heat to make them work.

Steam Engines

Steam is an excellent working substance for a heat engine. Heat is supplied to a steam engine by introducing the steam into the cylinder. (Fig. 7.10.) Although they can be used with the cheaper forms of fuel, such as coal, reciprocating steam engines are fast disappearing. Their efficiency is low, and they have a lower power-to-weight ratio than other engines.

High pressure steam

Low pressure steam

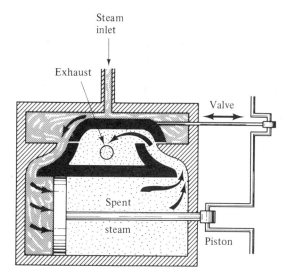

Fig. 7.10 Double acting steam engine. The valve is moved at the end of each stroke to direct the steam into the other half of the cylinder

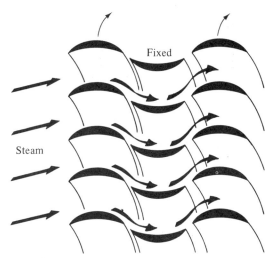

Fig. 7.11 Steam turbine showing the detail of the blades and the overall layout of a turbine

The frequent changes in the direction of a piston in a reciprocating engine cause both energy loss, due to friction, and vibration. These are reduced considerably in the steam turbine which can rotate at very high speeds. (7.11.)

Steam turbines operate at steam temperatures of 500°C and pressures of 200 atm and achieve actual thermal efficiencies of 35%. (Table 7.4.)

The Internal Combustion Engine

Few events in history have changed man's life so much as the development of the motor car. The internal combustion piston engine has played an important part in the change. There were early types of steam motor cars which had quite amazing performances, but the internal combustion engine has a higher power to weight ratio than the steam engine, which makes it more convenient for motor vehicles.

In the four-stroke engine, an explosive mixture

Table 7.4   Typical Thermal Efficiencies

Engine	η %
Turbojet aircraft engine	15
Petrol engine	15–20
Piston steam engine	12–27
Diesel engine	30–40
Steam turbine	30–50

of air and petrol vapour is drawn into the cylinder through the inlet valve. (Fig. 7.12.) The valve closes and the mixture is compressed as the piston returns on its next stroke. As the piston reaches the top of its stroke ignition is caused by an electric spark, and the mixture burns and expands rapidly, pushing back the piston. On the fourth stroke, the outlet valve opens and the burnt gas is pushed out by the piston through the exhaust pipe.

Notice that the thrust is only delivered on one stroke ($\frac{1}{2}$ revolution) while energy is absorbed on

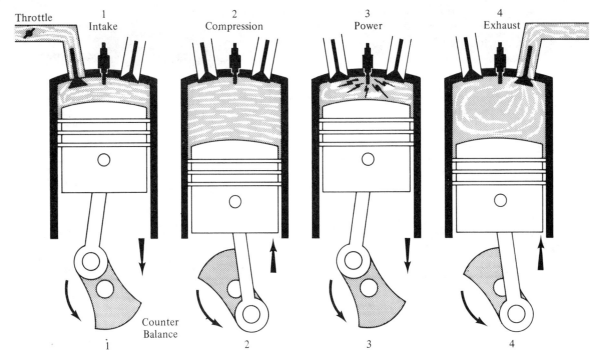

1 Intake   2 Compression   3 Power   4 Exhaust

Counter Balance

1   2   3   4

Fig. 7.12 The four-stroke internal combustion engine

the other three strokes ($1\frac{1}{2}$ revolutions). This gives the crankshaft an irregular motion which can be reduced, to some extent, by a massive flywheel. A more even drive is obtained by using several pistons on the same crankshaft.

Compression Ratio

Another important factor of the four-stroke engine is its compression ratio, that is, the ratio of the maximum and minimum volumes enclosed by the piston in its extreme positions. During the compression stroke, the fuel and air mixture gets hot. The greater the compression ratio, the higher the temperature becomes, and the greater the efficiency of the engine. A limit is reached when the temperature becomes high enough to ignite the mixture without the help of the spark. The pre-ignition of the charge may occur while the piston is still rising thus throwing great stress on the engine bearing and reducing the efficiency.

The Diesel Engine

A variation of the internal combustion engine combines the efficiency of a high compression ratio with the economy of using low grade fuel. In the diesel engine, air alone is drawn into the cylinder on the first stroke and compressed on the second stroke. Because the compression ratio is of the order of 16:1, the air is raised to a high temperature. Instead of a spark triggering combustion, a measured quantity of fuel is injected under pressure into the cylinder at a controlled rate. This ignites rapidly to produce the power stroke.

The efficiency of the diesel engine is greater than that of the petrol engine and, since it operates on cheaper fuel, it is far more economical. The diesel engine must be designed to withstand the high pressures and is, therefore, heavier and more expensive than the petrol engine. Diesels are preferred for heavy motor vehicles where economy and long life are the more important qualities.

The Wankel Engine

The pistons in a reciprocating engine may reach a speed of 100 m.p.h. and reverse their direction up to 100 times per second. These rapid reversals produce friction and wear additional to those produced by the load. Many attempts have been

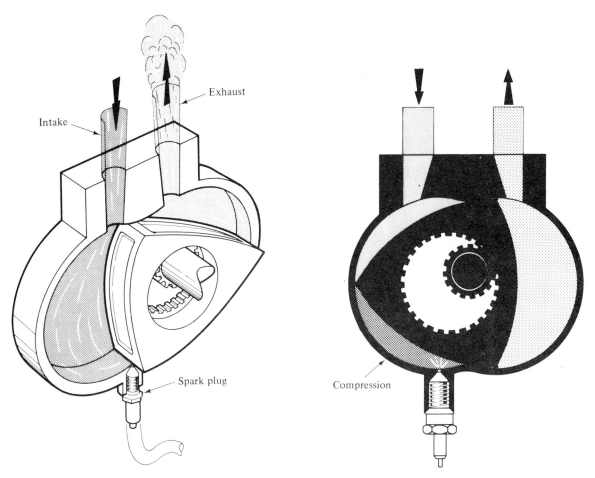

Intake

Exhaust

Spark plug

Compression

Fig. 7.13   The N.S.U.-Wankel rotary engine

made to produce a rotary gas engine but sealing problems have proved a great obstacle.

In the NSU-Wankel engine a rotor, shaped like an equilateral triangle with curved sides, rotates eccentrically in a rigid casing. (Fig. 7.13.) The casing is shaped rather like a circle with a waist. More exactly the case shape is a double epitrochoid and the corners of the rotor touch the inside of it in any position. The sides of the rotor enclose three spaces, which vary in volume as it rotates. During a complete revolution, each space goes through the four processes of the piston engine. The two rotor positions in Fig. 7.13 indicate six different stages in the complete cycle.

Aviation Engines

Two requirements of an aircraft engine are high power to weight ratio and a low air resistance. The petrol engine driving a propeller was at one stage used exclusively and it attained an output of 1.5 kW/kg weight. Its frontal resistance is high, however, and it loses its efficiency at high altitudes.

The disturbance produced in the air by the engine and the propeller produces a drag on the plane which cancels out some of the power of the engine. At high altitudes, the engine loses efficiency as the pressure of the air needed for combustion reduces.

Jet Engines

The turbojet engine (Fig. 7.14) was developed to meet the need for a high speed high altitude engine. Air is drawn into the open front of the engine, compressed, mixed with fuel, and ignited

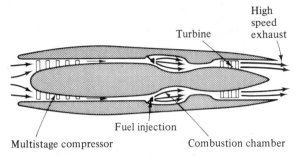

Fig. 7.14 Turbojet engine

in the combustion chambers. The high pressure exhaust from this chamber drives a gas turbine, which is coupled to the compressor by a central shaft, and then escapes at high speed at the rear of the engine. The turbine serves only to compress the incoming air while the propulsive force arises as a result of the reverse momentum of the exhaust material.

The advantage of the turbojet is its high power to weight ratio, which may be three times that of a petrol engine. The turbojet is most efficient between the speeds of 500 km/h and 1500 km/h, and it can operate at the reduced atmospheric pressure up to 20 km.

The Rocket Engine

A rocket engine is designed to operate in the absence of any atmosphere, and must carry both fuel and oxygen. (Fig. 7.15.) The oxygen and fuel, both in liquid form, are pumped into the combustion chamber where they combine to produce a high speed exhaust.

When the fuel of a rocket has been used up, the fuel tanks become so much dead weight. Rockets are therefore built in several stages so that the tanks can be jettisoned as soon as they are empty.

The Refrigerator

A refrigerator makes heat flow from a cold to a hot region contrary to its natural tendency. The second law of thermodynamics tells us that this cannot be done without the expenditure of work.

In the most common form of refrigerator, a volatile liquid such as ammonia or sulphur dioxide is allowed to evaporate as it flows through pipes surrounding the freezing compartment. (Fig. 7.16.) This extracts the latent heat of vaporization from the liquid which cools to a low temperature and thereby cools the interior of the refrigerator. The vapour at low pressure is drawn into a pump, which compresses it. The work done by the pump in compressing the vapour is transferred to the vapour, which is raised to a temperature above that of the room. As this vapour slowly moves through the cooling pipes, it is cooled below its boiling point, and returns to the liquid state. The cooling is sometimes accelerated by an electrically driven fan blowing air over fins on the cooling pipe. A valve controls the flow of liquid to the low pressure section, where it evaporates, and the cycle is repeated.

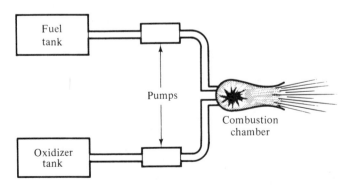

Fig. 7.15 The very high altitude rocket engine must be provided with both fuel and oxidizer

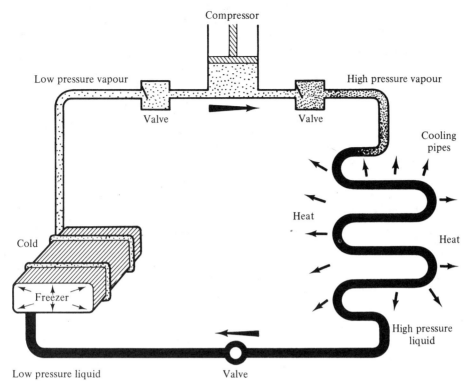

Fig. 7.16 **Refrigerator. Evaporation occurs as the working substance passes over the freezer**

Questions

1. Tractors run well on edible oil. Would you expect them to be more efficient than animals as machines?

2. What is the source of energy of (a) fossil fuels, (b) tides, (c) nuclear power?

3. What would be the effect on the world's climate if the specific and latent heats of water were about one tenth of their actual values?

4. How do latent heats of fusion and vaporization help in calibrating thermometers?

5. What property of water allows it to freeze on the surface while the body of the liquid is several degrees above freezing?

Multiple Choice

1. Which material would have a zero calorific value?
(a) wood, (b) sugar, (c) carbon, (d) carbon dioxide.

2. The unit of energy is the
(a) newton, (b) kilogram, (c) joule, (d) watt.

3. Specific heat capacity is measured in
(a) J, (b) J/K, (c) J/kg, (d) J/kg K.

4. Specific latent heat is measured in
(a) kg/J, (b) J, (c) J/kg, (d) W.

5. If 2.0 kg of copper (specific heat capacity 400 J/kg K) cools from 110°C to 100°C how much heat is lost?
(a) 20 J, (b) 800 J, (c) 4000 J, (d) 8000 J.

6. Which processes involve a change of state?
(a) condensation, (b) solidification, (c) conduction, (d) evaporation.

7. During evaporation the liquid remaining suffers a decrease in
(a) mass, (b) density, (c) temperature, (d) heat.

8. Which are reciprocating engines?
(a) turbine, (b) turbojet, (c) Wankel, (d) diesel.

9. Steam at 100°C burns the skin more than boiling water because it
(a) releases latent heat, (b) is a gas,
(c) has a higher temperature,
(d) has a higher specific heat.

10. Engine efficiency means

(a) $\dfrac{\text{work done}}{\text{time}}$, (b) $\dfrac{\text{work done}}{\text{energy supplied}}$, (c) $\dfrac{\text{heat used}}{\text{heat lost}}$,

(d) $\dfrac{\text{heat lost}}{\text{work done}}$.

Problems

1. How much heat is released by the combustion of 4.0 kg of petrol of calorific value 46×10^6 J/kg? [184×10^6 J]

2. How many joules of energy are produced by a 2 kW heater operating for 1 hour? [7.2 MJ]

3. How much heat is needed to heat 4.0 kg of lead from 20°C to 100°C? (Specific heat capacity of lead = 126 J/kg K.) [40.3 kJ]

4. What change in temperature is produced by 400 J of heat added to 100 g of (a) mercury, (b) iron, (c) wood? (Specific heat capacities: mercury = 126 J/kg K; iron = 480 J/kg K; wood = 1750 J/kg K.) [31.7 K, 8.3 K, 2.3 K]

5. If lead shot at 100°C is poured into 1 litre of water at 25°C, how much is required to raise the temperature to 50°C? (Specific heat capacities: lead = 126 J/kg K; water = 4200 J/kg K.) [16.7 kg]

6. A litre of boiling water is poured into an aluminium teapot at 20°C. If the final temperature of water and teapot is 86°C, what is the mass of the teapot? (Specific heat capacities: aluminium = 890 J/kg K; water = 4200 J/kg K.) [1.0 kg]

7. By how much is the temperature of water raised in going over Niagara Falls, 51 m high? ($g = 9.8$ m/s².) (Specific heat capacity of water = 4200 J/kg K.) [0.12 K]

8. What is the upper limit to the power of an engine burning 18 kg of diesel oil per hour, if the calorific value of the oil is 50 MJ/kg? [250 kW]

9. An engine burns 35 kg of fuel oil of calorific value 40×10^6 J/kg and performs 4.4×10^8 J of work. Calculate its efficiency. [31%]

10. Freon, which is often used in refrigerators, has a specific latent heat of vaporization of 168 kJ/kg. How much freon is vaporized if 336 kJ of heat are transferred from the storage compartment of a refrigerator? [2.0 kg]

11. A boiler delivers steam at atmospheric pressure to a radiator, where it condenses and cools to 80°C before returning to the boiler. How much steam must be circulated to supply 9.4×10^6 J of heat? Obtain data from Tables 7.2 and 7.3. [4.0 kg]

12. A heater supplying 168 W is lowered into 200 g of water at 20°C and removed after 14 min. If 30 g of water have boiled away, what is the specific latent heat of vaporization of water? Specific heat capacity of water = 4200 J/kg K. [2.5×10^6 J/kg]

13. What is the least amount of heat that will melt 6.0 kg of lead, initially at 20°C? (Specific latent heat of fusion of lead = 24.7 kJ/kg; specific heat capacity of lead = 1.26×10^2 J/kg K; melting point of lead = 327°C.) [380 kJ]

14. What is the result of removing 3000 kJ from 1.0 kg of steam at 100°C? Obtain data from Tables 7.2 and 7.3. [0.96 kg of ice and 0.04 kg of water at 0°C]

Multiple Choice Answers

1(d), **2**(c), **3**(d), **4**(c), **5**(d), **6**(abd), **7**(acd), **8**(d), **9**(a), **10**(b).

8.
Heat Transfer

Everywhere in Nature heat energy is on the move. The heat moves from high to low temperature regions tending to produce a uniform temperature throughout the universe. On the scale of the universe, heat transfer occurs slowly, and temperature differences will exist for all the foreseeable future. Viewed on a smaller scale heat transfer can be violently rapid, as in a nuclear explosion.

The control of heat flow has always been of importance to man. From the most primitive times, he has needed to protect his body from the extremes of heat and cold. Even today this remains his first consideration but, he now wishes to control the transfer of heat in a far wider variety of applications, from the incubation of bacteria to atomic energy reactors. We recognize three basic processes of heat transfer: conduction, convection, and radiation. The process of evaporation often makes an important contribution and we could well include it in the list. The human body, in particular, relies on evaporation to keep its temperature down in hot climates.

In some instances, we try to promote heat flow as in a boiler or heat exchanger while, in other cases, we try to reduce heat flow as in homes in winter. Sometimes we delicately control the heat flow both in and out as in a manned satellite. In practice, engineers solve heat transfer problems by analyzing the transfer into its component processes and dealing with each process separately. We will adopt the same approach and consider each of the processes in turn.

Conduction

In conduction, heat flows down temperature gradients from places of high temperature to places of lower temperature. When heat has been passing through a conductor for some time, temperatures along the conductor may reach steady values.

In a bar which is lagged at the sides, the heat flows longitudinally in at one end and out at the other. The temperature gradient is uniform as shown in Fig. 8.1(a). This distribution of temperature would also be typical of a conductor of large area, such as a boiler plate.

In a conductor which is not lagged at the sides, the temperature quickly falls away from the hot end, owing to the heat escaping through the sides. (Fig. 8.1(b).) Such a distribution would apply to the case of the cooling fins on an air-cooled engine. The optimum length of the fins may be deduced from a knowledge of the temperature distribution along the fins.

Thermal Conductivity *k*

The rate of heat conduction through a material is measured in joules per second, i.e., watts. The rate depends on three factors. (Fig. 8.2.)

(a) The temperature gradient $\dfrac{T_1 - T_2}{x}$

(b) The area A
(c) The material

If a quantity of heat Q flows during a total time t, then:

$$\text{Rate of flow} \quad \frac{Q}{t} \propto \frac{A(T_1 - T_2)}{x}$$

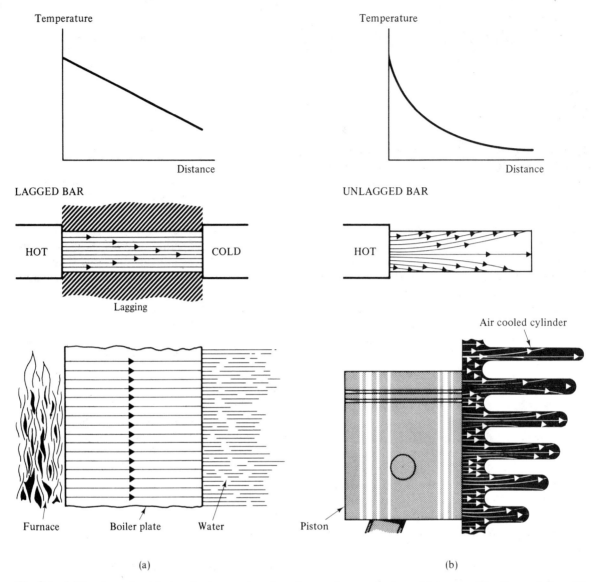

Fig. 8.1 (a) In a lagged conductor, the temperature drops by equal amounts at equal intervals along the conductor. (b) In an unlagged conductor, the heat flow is not parallel and the temperature gradient is not linear

If we introduce a constant of proportionality k this will depend on the material.

$$\frac{Q}{c} = kA \frac{(T_1 - T_2)}{x}$$

The constant k is called the thermal conductivity of the material and is defined quite adequately by this equation. Alternatively, it could be defined in words as **'the rate of flow of heat through the material, per square metre per unit temperature gradient'.** The units of thermal conductivity are watts per metre kelvin (W/m K).

Problem

An icebox has an area of $2.4 \, \text{m}^2$ and its inner surface is maintained at $2°C$ when the outer surface is at $17°C$. If the box is insulated with cork 50 mm thick, how much heat flows into the box per hour? Take k for cork to be $0.046 \, \text{W/m K}$.

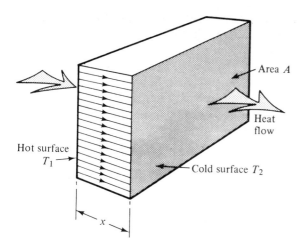

Fig. 8.2 Factors involved in the heat flow through a section of material

Table 8.1

The figures should be taken as approximate because of variations of k with temperature and with the particular state of the material

Material	Thermal conductivity, W/m K
Copper	400
Aluminium	210
Iron	67
Glass	1.1
Concrete	0.92
Brick	0.63
Water	0.59
Alcohol	0.21
Wood	0.17
Hydrogen	0.17
Cork	0.046
Wool felt	0.042
Air	0.025

Solution

$$Q/t = kA \times \text{temperature gradient}$$
$$= 0.046 \times 2.4 \times (17-2)/0.05 \text{ J/second}$$
$$= 60 \times 60 \times 0.046 \times 2.4 \times 15/0.05 \text{ J/hour}$$

Flow of heat
$$= 10.8 \times 10^4 \text{ J/hour}$$

Problem

The plaster ceiling of a centrally heated house is 6 m × 10 m × 12 mm thick. Calculate the heat flow through the ceiling if the inside and outside surfaces are at a temperature of 24°C and 4°C. (k for plaster = 0.63 W/m K.)

Solution

Flow through uninsulated ceiling

$$Q/t = kA\theta$$

$$\text{Rate of flow} = \frac{0.63 \times 6 \times 10 \times (24-4)}{0.012}$$

$$= 6.3 \times 10^3 \text{ J/s} = 6.3 \text{ kW}$$

Rate of heat loss = 6.3 kW

(Why is this figure so ridiculously high?)

U Values

In practice, when heat is flowing through a solid surface which is in contact with a liquid or gas, a large part of the temperature change occurs in the fluid near the surface. These surface layers strongly affect the rate of conduction through a barrier such as a boiler plate or a window. (Fig. 8.3.) If the heat flow is calculated on the assumption that the temperatures of the surfaces of the barrier are the same as those in the body of the fluids, it could be many times too great.

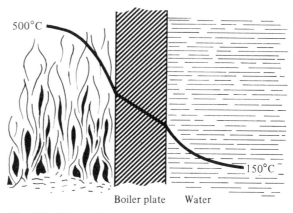

Fig. 8.3 Most of the temperature drop at a boiler plate is in the fluid layers near its surface

Where we are dealing with standard structures such as a brick wall or plate glass, it is more accurate to give an experimentally measured figure for the structure as a whole, rather than to quote the conductivities of its materials. (Table 8.2.) Such a figure is called a U value and is defined as the rate of flow of heat per unit area, per degree of difference of temperature across the structure.

Rate of flow of heat = U × area × temp. difference

Table 8.2

Structure	U value, W/m² K
Solid double brick wall	2.8
Solid double brick wall, plastered one side	2.5
Plate glass (6 mm)	10
Plastered ceiling	2.0

Problem

Calculate the U value of the uninsulated ceiling in the previous example if its U value is 2.0 W/m² K.

$$\text{Heat loss} = U \times A \times \theta$$
$$= 2 \times 6 \times 10 \times (24-4)$$
$$\text{Heat loss} = 2400 \text{ W}$$

Measurement of Thermal Conductivity

To determine the thermal conductivity k of a material, we must measure the quantities in the equation:

$$k = \frac{\text{rate of heat flow}}{\text{area} \times \text{temperature gradient}}$$

Because the temperature gradients in good conductors are small, they are difficult to measure accurately. This is overcome in Searle's apparatus, by using a test specimen about 40 cm long which gives an appreciable change in temperature at two widely separated points. (Fig. 8.4.) A regular flow of water through a spiral coil, which is soldered on to the bar, extracts the heat from the cold end of the bar. We measure the rise in temperature of this water and its rate of flow, and calculate the rate of conduction of heat along the bar. Apart from demonstration apparatus, the measurement is made only in specialized laboratories because the apparatus must be constructed around the specimen.

Heat flow per second
$$= Q/t = \text{mass of water per second}$$
$$\times \text{specific heat capacity of water}$$
$$\times \text{temperature difference}$$
$$= (\text{mass of water}/t) \times 4.2 \times (T_4 - T_3)$$
Temperature gradient $= (T_1 - T_2)/L$
Thermal conductivity

$$k = \frac{(\text{mass of water}/t) \times 4.2 \times (T_4 - T_3)}{A(T_1 - T_2)/L}$$

Conduction in Metals

Metals are very good thermal conductors, just as they are good electrical conductors. We can, in fact, say more than this because measurements show that the ratio of the thermal to the electrical

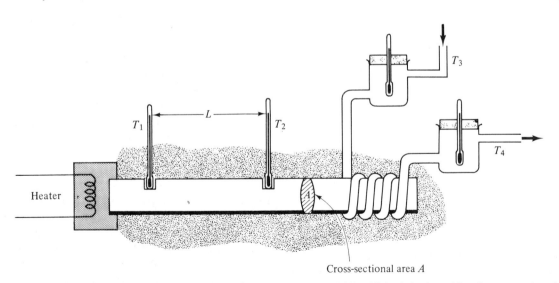

Cross-sectional area A

Fig. 8.4 Searle's apparatus. Heat is prevented from escaping through the sides of the bar either by evacuating the space around it or by filling the space with insulating material

conductivity is a constant for metals at the same temperature.

$$\frac{\text{Thermal conductivity}}{\text{Electrical conductivity}} = Z$$

The simplicity of this relation is not just a coincidence but suggests some underlying similarity between the processes. What is behind this similarity? Electrical current in metals is known to be a flow of electrons. The electrical conductivity is high because in metals there are large numbers of electrons not rigidly bound to particular atoms. We strongly suspect that these same highly mobile electrons are responsible for passing on the heat so quickly by their rapid reaction on each other and on the lattice of metal atoms.

It appears that conduction proceeds by two mechanisms, one involving the atoms, and the other a much more rapid one involving any loosely bound electrons which may be present.

One of the best conductors is the metal sodium which is known to have electrons which are highly mobile. Unfortunately, it is also highly chemically reactive, and ignites spontaneously even at normal temperatures, which makes it difficult to handle. We use it as a conducting material in special cases where it can be enclosed, for example, in the valves of certain internal combustion engines. (Fig. 8.5.)

The group of metals, silver, copper, and gold combine high electrical and thermal conductivity with great resistance to corrosion, but unfortunately only copper is cheap enough to be widely used.

Conduction in Non-Metals

Insulators are generally materials which contain comparatively few free electrons, and heat flows entirely by means of the interaction of the atoms. In comparison with electrons, these atoms are massive, and the transfer of heat by this process is usually very much slower.

The atoms have a vibratory motion about an equilibrium position but each atom keeps its place in the lattice. Atoms at a higher temperature have more energetic vibrations and, by jostling their neighbours, cause them to vibrate more energetically. They, in turn, influence other atoms and so the energy is passed on through the material. When the temperatures have stabilized, the molecules at a particular point on the conductor will maintain the same degree of vibration, being simultaneously activated by molecules in the direction of higher temperature and damped to the same extent by the molecules in the direction of lower temperatures. (Fig. 8.6.)

Convection

Convection is the transfer of heat in which the material moves taking the heat with it. Liquids and gases, which are poor conductors, can transfer heat very effectively by this means. Natural convection occurs because of the reduction in density which usually accompanies heating. By the principal of flotation, the hotter (lighter) material rises displacing or mixing with the cooler material above. (Fig. 8.7.) Household hot water systems use natural convection to circulate the water. Chimneys promote convection by insulating the hot gas from the cooler atmosphere.

Where natural convection is too slow for a certain purpose, the fluid velocity may be increased by mechanical means such as a fan or pump. We call this forced convection and use it to produce a more rapid replacement of the fluid in contact with the heat source. It allows apparatus

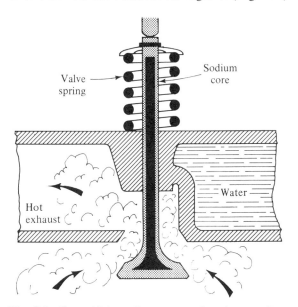

Fig. 8.5 Some high performance engines have sodium cored valves to conduct the heat away rapidly

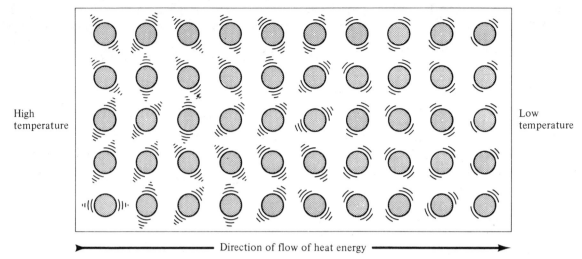

Fig. 8.6 Molecules at a certain point in a material receive energy from the molecules on one side and give up energy to the molecules on the other side

to be smaller than would be necessary using natural convection. The small bore central heating system is an example of the use of forced convection in this way, while the cooling system of a car engine uses forced convection of water and air. (Fig. 8.8.)

The effectiveness of convection depends very largely on the viscosity of the fluid concerned. Thus, thin soup can be left to simmer in the pan but porridge will surely burn if not stirred.

Conduction in Fluids

Since in most liquids the process of heat conduction, as distinct from convection, is rather sluggish, it is thought that conduction in liquids is similar to that in solid insulators. There are some notable exceptions such as mercury and other molten metals. The Dounreay Fast Reactor for producing atomic power uses a liquid sodium-potassium alloy as its coolant.

Gases, too, are generally good insulators. Heat travels in them rather differently from the way it travels in solids and liquids. The highly mobile gas

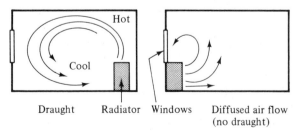

Fig. 8.7 Natural convection in a room may cause a draught if the 'radiators' are not well situated

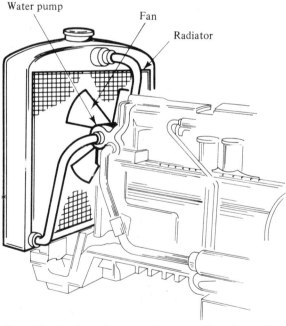

Fig. 8.8 To keep it as small as possible, the cooling system of a car engine uses forced convection of water and air

molecules convey heat by the process of convection. In practice, we find it difficult to separate the effect of convection from our examination of conduction in gases. At very low pressures, the molecules of gas are so scattered that they rarely collide, and convection and conduction become indistinguishable. Many solid structures, such as cork, wool, and expanded polystyrene, act as good insulators because of the pockets of still air trapped inside them.

Evaporation Transfer

A further means of heat transfer is that of evaporation and its reverse process condensation. Whenever exposed liquid surfaces are involved, evaporation may occur and extract the latent heat of vaporization from the remaining liquid. If, on the other hand, condensation occurs, this latent heat is given up at the condensing surface. The effect of evaporation is likely to be very pronounced when the liquid involved is volatile, or is near to its boiling point.

A boiling kettle is a good example of evaporation and condensation transfer. The source of heat provides the latent heat to change the water into vapour, while the temperature remains constant. The vapour diffuses away from the kettle until it encounters perhaps the cold surface of a wall or window. Here it condenses to a liquid, at the same time giving its latent heat to the wall. The net result has been a rapid transfer of heat from the source to the wall. Unfortunately for some wall coatings, the water has also been transferred.

The excess heat from the human body is effectively dissipated by evaporation. Each gramme of water evaporated extracts about 2000 joules, and in a hot climate a man may sweat away 3 kg of water per day.

Radiation

Molecules in a surface vibrating with thermal energies generate electromagnetic waves. The waves travel away at the speed of light, taking energy with them and leaving the surface cooler.

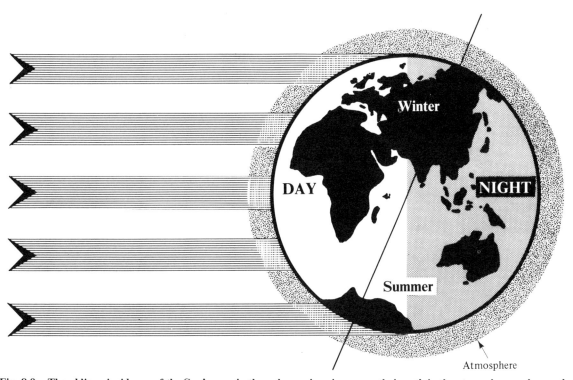

Fig. 8.9 The oblique incidence of the Sun's rays in the polar regions increases their path in the atmosphere and spreads them more thinly over the surface

When the waves encounter some other material, they produce a disturbance of the surface molecules, thus raising their temperature. The radiation requires no intervening material between transmitter and receiver and, in fact, travels best in a vacuum.

Radiation is particularly important because it is the process by which heat is received by the earth from the sun. When the sun is directly overhead it provides about 1 kW/m^2. The level of radiation is less in polar regions of the earth because it meets the surface obliquely and traverses a longer path in the atmosphere. (Fig. 8.9.) A similar level of radiation is received by the moon's surface but, because the moon revolves only once in four weeks, each point on the moon enjoys two weeks sunshine followed by two weeks of darkness. The temperature extremes produced on the moon by this long day and night are much wider than on earth. This causes one of the major problems for manned landings. The solution may not be in elaborate insulated suits but in keeping to the twilight zones where the temperature is moderate.

Nature of Heat Radiation

Heat radiation contains a large spread of wavelengths, the average value of the wavelength emit-ted depending on the temperature of the radiating surface. The larger part of the energy in heat radiation has a wavelength falling in the infra-red region of the complete electromagnetic spectrum. At higher temperatures, about 1000 K, the proportion of shorter wavelength radiation increases sufficiently to be seen by the eye. Because infra-red radiation and light are both electromagnetic waves, they are reflected and refracted in a similar way. (Figs. 8.10 and 8.11.)

The materials of prisms and lenses used to examine heat radiation must be carefully chosen to transmit the wavelengths required. For example, glass is transparent only to light and the shorter wavelengths of infra-red radiation. It is opaque to the longer wavelength components. Quartz, on the other hand, is transparent to a wider range of wavelengths extending far into the infra-red.

The cold frame or greenhouse is a clever means of trapping heat radiation. The radiation from the sun at a temperature of 6000 K contains a large proportion of short wavelength radiation which is admitted by the glass. (Fig. 8.12.) The interior of the greenhouse at a temperature of 300 K radiates longer wavelength radiation, to which the glass is more opaque. Thus, the temperature of the interior rises considerably above that of the surrounding atmosphere.

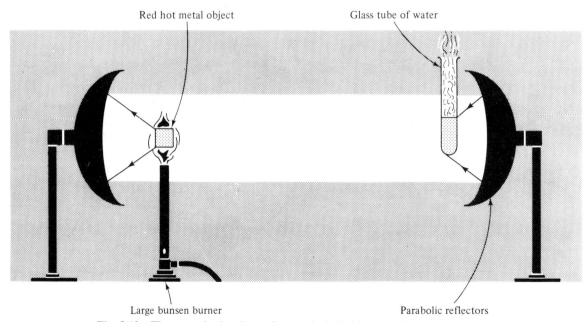

Red hot metal object   Glass tube of water

Large bunsen burner   Parabolic reflectors

Fig. 8.10 The water in the glass tube may be boiled by the reflected radiant heat

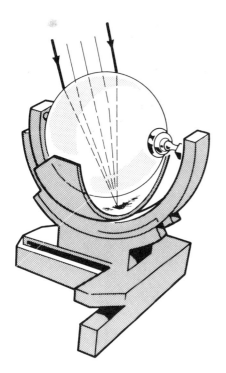

Fig. 8.11 In a sunshine recorder, a glass sphere focuses the heat radiation which scorches a line on a paper during the hours of sunshine. Glass strongly absorbs infra-red radiation, so to examine the radiation in detail, we use instruments which have quartz or rock salt components

6000 K
The wave-lengths from the sun are mostly short and pass through the glass
300 K
The longer wave-length radiation from the greenhouse interior is blocked by the glass

Fig. 8.12 Glass acts as a one way trap for the sun's radiant energy

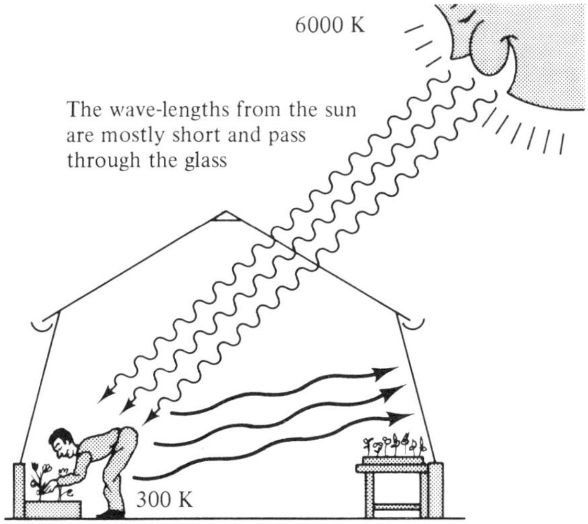

Effects of Nature of Surface

The rate of radiation of heat from a body depends on the nature of its surface. A polished metal surface is a poor radiator, while a white painted surface is only slightly better. Radiation increases as the shade of the surface darkens and loses its shine, until a matt black surface is the best radiator of all. (Fig. 8.13.) If these surfaces are exposed to radiation, it is found that the best radiating surface, the matt black one, is also the most effective absorber. The same order is preserved up to the polished metal surface which absorbs least. Polished metal surfaces reflect heat best and they are used on electric fires and the outside of satellites. To keep out the heat in the tropics, people wear white clothes and live in white-washed houses. The so-called household radiators should be matt black for maximum heat transfer but people prefer to paint them to match their walls. This makes little difference because most of the heat from these 'radiators' is lost by convection.

Cooling Laws

All bodies at any temperature above absolute zero radiate heat energy. Most of the energy is in the infra-red radiation which we can feel but not see. Stefan made some careful measurements and

Shiny black
White
Hot water
Pyrometer
Matt black
Shiny metal

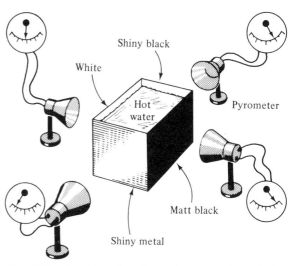

Fig. 8.13 Leslie's cube allows the comparison of the radiant energy from four different surfaces at the same temperature

found that the radiation is proportional to the fourth power of the absolute temperature.

$$\text{Energy lost per second} \propto T^4$$

This is the Stefan law of radiation. We can write

$$\text{Energy loss} = \sigma \times \text{area} \times T^4$$

σ is the Stefan constant $= 5.7 \times 10^{-8}$ W/m^2 K^4

At the surface of the sun (6000 K), the flow of radiation energy is colossal: $\propto (6000)^4$. In the interior of stars, like the sun, where the temperature is of the order of millions of degrees the radiation is so intense as to exert a substantial physical pressure. This radiation pressure opposes the gravitational contraction and is an important factor in the stability of stars.

Radiation does not, as one might expect, reduce the temperature of a body to absolute zero. A body surrounded by surfaces at the same temperature, absorbs and emits energy at the same rate, and there is no net change in temperature. When a body is hotter than its surroundings, it radiates more heat than it receives and its temperature drops.

For large temperature excesses, e.g., objects glowing red hot, the radiation from the surroundings is insignificant compared with the radiation from the hot object.

$$\text{Rate of cooling} \propto T^4$$

For small temperature excesses convection conduction and radiation all play a significant part and the rate of cooling is directly proportional to the excess temperature.

$$\text{Rate of cooling} \propto T_{\text{object}} - T_{\text{environment}}$$

This is Newton's Law of Cooling. It is an approximate law but a useful one. We use it to predict heat losses at various temperatures so that we can design suitable heating systems. Figures 8.14 and 8.15 show how a knowledge of heat transfer processes has been applied to insulation problems. As fuel becomes more expensive good insulation of buildings is essential and saves its cost in a short time. The main routes for heat loss in a house are excessive ventilation (draughts) and conduction through the roof.

86

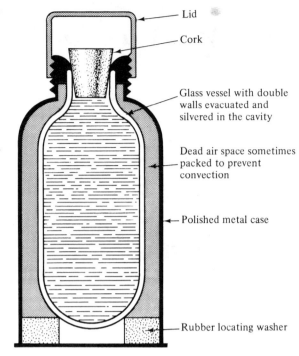

Fig. 8.14 The vacuum flask is a container which is designed to minimize heat flow either in or out

Labels: Lid; Cork; Glass vessel with double walls evacuated and silvered in the cavity; Dead air space sometimes packed to prevent convection; Polished metal case; Rubber locating washer

Problem
It costs £6 per week to maintain the temperature of a house at 20°C when the outside temperature is at 5°C. What would it cost per week to maintain the temperature at 25°C?

Solution

Rate heat loss $\propto$ excess temperature
At 20°, heat loss $\propto 15°$
At 25°, heat loss $\propto 20°$

Hence cost $= £6 \times \dfrac{20}{15} = £7.50$ per week

Problem
The filament of an evacuated light bulb requires 40 W to maintain it at 2000 K. How many watts would be needed to maintain it at 2200 K, assuming that the heat lost by conduction is negligible?

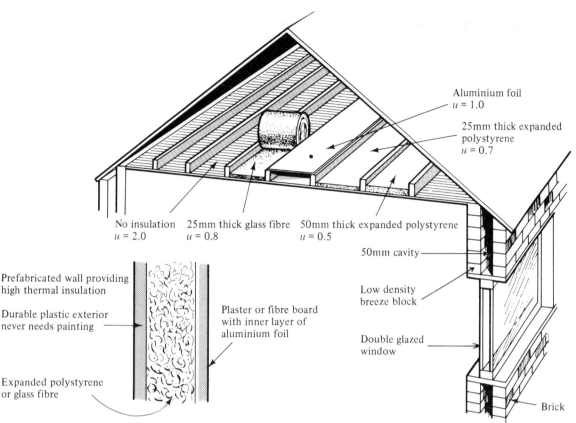

Fig. 8.15 Heat insulation in a building

Solution

Rate of radiation $\propto T^4$

At the higher temperature $= \dfrac{2200^4}{2000^4} = 1.46$

heat losses increase in ratio.

Hence power required $= 40 \times 1.46$

Power required $= 58.4$ W

Questions

1. If evaporation cools, does condensation warm?

2. Why is steel wool a worse conductor than the steel from which it is made?

3. Why does a metal spoon at $0°C$ feel colder to the touch than a wooden one at the same temperature?

4. Why does damp air feel warmer than dry air at the same temperature?

5. Why does a cube cool quicker than a sphere of the same mass and material under identical conditions?

Multiple Choice

1. Heat losses by conduction from an uninsulated house are greatest through the
   (a) windows, (b) roof, (c) walls, (d) floor.

2. Which process involves transferring matter?
   (a) radiation, (b) conduction, (c) convection, (d) illumination.

3. Which process transfers heat through a vacuum?
   (a) conduction, (b) convection, (c) radiation, (d) condensation.

4. Which is a good heat conductor?
   (a) air, (b) cork, (c) water, (d) zinc.

5. A solid at a temperature of 800 K will lose most of its heat by
   (a) conduction, (b) convection, (c) radiation, (d) evaporation.

6. In a hot climate the human body is cooled by the process of
   (a) conduction, (b) convection, (c) evaporation, (d) radiation.

7. In a cold climate body heat is lost mainly by
   (a) conduction, (b) convection, (c) radiation,
   (d) evaporation.

8. Heat losses due to convection are increased
   (a) in still air, (b) in a wind, (c) by lagging,
   (d) for polished surfaces.

9. Heat radiation losses are greatest for which of these types of surface?
   (a) white, (b) shiny, (c) metallic, (d) matt black.

10. Heat conduction through a barrier is independent of
    (a) temperature difference, (b) colour, (c) thickness,
    (d) thermal conductivity.

Problems

1. Calculate the rate of heat flow through a brick wall 0.20 m thick measuring 3.0 m × 4.0 m if the temperature of its surfaces are 0°C and 15°C. Thermal conductivity = 0.63 W/m K. [567 W]

2. Calculate the rate of heat flow through a 2.0 m × 1.0 m sheet of glass 5.0 mm thick when there is a temperature difference between its surfaces of 1.2 K. Thermal conductivity of glass 1.1 W/m K. [528 W]

3. The glass in the windows of a house has a total area of 20 m². If the glass is 4.0 mm thick and the temperature differences between its two surfaces is 1 K, how much heat is transmitted in 24 hours? (Thermal conductivity of glass = 1.1 W/m K.) [4.7×10^8 J]

4. If the earth, 150 000 000 km from the sun, receives 1400 J/m² s, how much heat is radiated by the sun per second? [3.96×10^{26} J]

5. A blackened spherical retort at 727°C has a diameter of 1.2 m. At what rate must heat be supplied to balance radiation losses? (Stefan constant = 5.7 × 10^{-8} W/m² K⁴.) [2.57×10^5 W]

6. The ceiling of a building measuring 9.0 m × 7.5 m is insulated with fibreglass. The temperature below the ceiling is 30°C and above it is 0°C. What is the rate of heat loss through the ceiling? (U value = 0.6 W/m² K.) [1.2 kW]

7. A filament is supplied with 100 watts which maintains its temperature at 1200°C. To what temperature would it rise if it were supplied with 150 watts? [1360°C]

8. What is the temperature difference across a glass window 4.0 mm thick if 250 W escape per square metre? How much would it cost per day to replace it at 6p per kilowatt-hour? (Thermal conductivity of glass = 1.1 W/m K.) [0.9 K; 36p]

Multiple Choice Answers

1(b), 2(c), 3(c), 4(d) 5(c), 6(c), 7(b), 8(b), 9(d), 10(b).

9.
Atomic Structure

An apparently endless variety of materials exists in nature. Add to these, the materials produced by man the number rises to a staggering total. Yet despite the diversity in the appearance of matter, men have always cherished the idea of an underlying simplicity. Long before science developed to the stage of experiment, the scientists of the ancient world speculated about the basic substances from which all matter was made, and formulated the theory that there were four elements—earth, water, fire and air. (Fig. 9.1.)

This theory obviously sprang from common human observation; if we interpret the four elements as solid, liquid, heat, and gas, they include the whole of man's ordinary experience. By means of this theory men succeeded for many centuries in

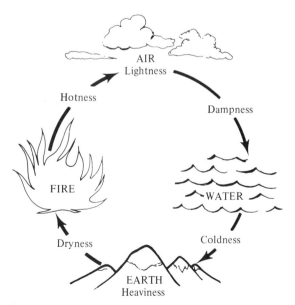

Fig. 9.1   The four element theory of the nature of matter

accounting to their own satisfaction for most natural phenomena. For example, they explained the fact that steam rises contrary to normal gravity by pointing out that steam is a mixture containing more fire than water, so that it shares with flames the property of rising. Differences in the appearance of materials, they said result from differences in the proportions of the four elements contained in them. All they felt they had to do to change one material into another was to find a way of altering the proportions of the constituent elements. By adding a sufficiently large dose of the element of fire to the correct mixture of sand, alkali, and lead, they produced glass, and for several wasteful centuries alchemists tried continually to change worthless materials into gold by similar processes of mixing and heating.

To us the speculative methods of the early scientists, mixing and stirring their quaint ingredients like so many adventurous cooks, and making new discoveries only once in a long while—and then only by accident—seem naive and amusing. Yet this was, in general, the state of science before the vigorous new approach, which we call the Scientific Revolution, began in Western Europe, about 400 years ago. Then, for the first time, scientists started using systematic experiments to seek out the underlying structure of matter. They slowly revealed the existence of, not 4, but about 100 elements, of which all other substances are composed.

The Elements

The elements which occur in nature range from hydrogen, which is the lightest, to uranium, which

89

is the heaviest. Rarely does any one of the elements occur separately from other elements. When an element does appear in its pure form it is usually to be found mixed with other matter, as gold-dust is with river-sand, and oxygen with the other gases in the atmosphere. More often, the elements only exist in chemical combination with other elements, and their isolation requires a series of chemical reactions. All the naturally occurring elements have been isolated and identified. They have been given names and symbols, as shown in the complete list given in Table 9.1.

Under normal conditions eleven of the 100 or so elements are gases. They include common gases like oxygen and nitrogen, pure gases like argon and radon, and poisonous gases like chlorine and fluorine. Only two of the elements are liquids: mercury and bromine. All the rest are solids.

The Atom

If it were possible to take a quantity of any one of the elements and to divide it into a large number of very small amounts, is there any limit to the size of the parts, or could we go on dividing indefinitely? The answer is that eventually we would come to a dead stop. We would reach a stage at which we could not divide the element any further.

Table 9.1   The Elements

At. no. (Z)	Symbol	Element	At. no. (Z)	Symbol	Element	At. no. (Z)	Symbol	Element
1	H	Hydrogen	36	Kr	Krypton	71	Lu	Lutecium
2	He	Helium	37	Rb	Rubidium	72	Hf	Hafnium
3	Li	Lithium	38	Sr	Strontium	73	Ta	Tantalum
4	Be	Beryllium	39	Yt	Yttrium	74	W	Tungsten
5	B	Boron	40	Zr	Zirconium	75	Re	Rhenium
6	C	Carbon	41	Cb	Niobium	76	Os	Osmium
7	N	Nitrogen	42	Mo	Molybdenum	77	Ir	Iridium
8	O	Oxygen	43	Tc	Technetium	78	Pt	Platinum
9	F	Fluorine	44	Ru	Ruthenium	79	Au	Gold
10	Ne	Neon	45	Rh	Rhodium	80	Hg	Mercury
11	Na	Sodium	46	Pd	Palladium	81	Tl	Thallium
12	Mg	Magnesium	47	Ag	Silver	82	Pb	Lead
13	Al	Aluminium	48	Cd	Cadmium	83	Bi	Bismuth
14	Si	Silicon	49	In	Indium	84	Po	Polonium
15	P	Phosphorus	50	Sn	Tin	85	At	Astatine
16	S	Sulphur	51	Sb	Antimony	86	Rn	Radon
17	Cl	Chlorine	52	Te	Tellurium	87	Fr	Francium
18	A	Argon	53	I	Iodine	88	Ra	Radium
19	K	Potassium	54	Xe	Xenon	89	Ac	Actinium
20	Ca	Calcium	55	Cs	Caesium	90	Th	Thorium
21	Sc	Scandium	56	Ba	Barium	91	Pa	Protactinium
22	Ti	Titanium	57	La	Lanthanum	92	U	Uranium
23	V	Vanadium	58	Ce	Cerium	93	Np	Neptunium
24	Cr	Chromium	59	Pr	Praseodymium	94	Pu	Plutonium
25	Mn	Manganese	60	Nd	Neodymium	95	Am	Americium
26	Fe	Iron	61	Pm	Promethium	96	Cm	Curium
27	Co	Cobalt	62	Sa	Samarium	97	Bk	Berkelium
28	Ni	Nickel	63	Eu	Europium	98	Cf	Californium
29	Cu	Copper	64	Gd	Gadolinium	99	Es	Einsteinium
30	Zn	Zinc	65	Tn	Terbium	100	Fm	Fermium
31	Ga	Gallium	66	Dy	Dysprosium	101	Md	Mendelevium
32	Ge	Germanium	67	Ho	Holmium	102	No	Nobelium
33	As	Arsenic	68	Er	Erbium	103	Lw	Lawrencium
34	Se	Selenium	69	Tm	Thulium			
35	Br	Bromine	70	Yb	Ytterbium			

Heating or cooling, chopping or cutting would make no permanent change in this basic particle. Each of the elements has such a basic unit called an atom. It is defined as the smallest part of an element which can take part in a chemical change.

The Molecule

When elements combine together, the atoms join to form a basic unit of the new substance. This unit is called a molecule. It is the smallest part of a substance, which can have a separate stable existence.

Even atoms of a single element may be more stable when they are joined together into a molecule. Oxygen atoms, for example, are more stable when they join together in pairs. (Fig. 9.2.) The molecule of oxygen is then said to be diatomic. It is represented by writing a suffix after the symbol for the element, e.g., O_2 for a molecule of oxygen, and H_2 for a molecule of hydrogen, which is also diatomic.

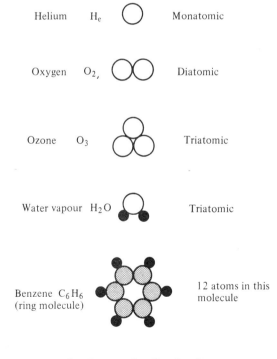

Helium H_e Monatomic

Oxygen O_2,   Diatomic

Ozone O_3 Triatomic

Water vapour H_2O Triatomic

Benzene C_6H_6 (ring molecule)   12 atoms in this molecule

Polyethylene (Polymer) $n(CH_2)$ Polyatomic

Fig. 9.2 Atoms combine together to form molecules

The atoms of many elements are quite stable when they exist singly and so they are both atom and molecule at the same time. Such elements are monatomic, e.g., Helium, He.

In some circumstances, oxygen atoms combine together to form a triatomic molecule, that is a molecule containing three atoms, which we represent by the symbol O_3. This form of oxygen is produced near electrical equipment where sparks occur. It has the special name of ozone, and can be distinguished from oxygen, which is odourless, by its smell. (Ozone is produced in the ionosphere of the earth, where cosmic rays from outer space produce an electrical disturbance as they encounter the earth's atmosphere.) Some atomic groups combine together to form chains of great length called polymers. (Fig. 9.2.) Such molecules are the basis of many plastics and synthetic textiles, and chemists can tailor molecules to give required properties.

The Mass of Atoms and Molecules

The mass of an atom expressed in kilograms is a very small quantity. For example, the mass of a carbon atom is 2×10^{-26} kg. A much smaller unit of mass is more convenient for expressing the masses of atoms. The hydrogen atom is the lightest, so that by taking its mass as one unit, we can express the mass of all the other atoms as numbers greater than one. This unit we call a unified atomic mass unit.

1 unified atomic mass unit
$$= 1 \text{ u} = 1.67 \times 10^{-27} \text{ kg}$$

On this scale the carbon atom's mass is 12.0 u. The mass of an atom expressed in atomic mass units is referred to as the relative atomic mass of the element. The atomic mass unit also serves as a unit for measuring the masses of molecules and, in this case, they are referred to as relative molecular masses.

The Structure of Atoms

Having established that all matter is composed of only about 100 different sorts of atoms, one might think that man's mind would rest content. His curiosity knows no bounds, however, and there arose the inevitable question: 'what are atoms made of?' Certain similarities in the chemical properties of different elements seemed to point to a similarity in structure. Step by step, by painstaking and inspired experiment, a picture of atomic structure was developed, and is still developing. The most adequate theories are highly mathematical, but a useful picture which appeals to common sense and yields a good measure of understanding of matter, is given by the orbiting electron model of the atom.

The Value of Models

You may wonder why we bother to explain at length a mental picture of the atom which is known to be ultimately inadequate. To answer this question we can look to the history of science which has developed by a series of concepts each one improving on the earlier ones. For example, the particle theory of light gave way to the wave theory, and the caloric theory of heat gave way to the kinetic theory. Even today simplified theories, although they may contain half truths, are useful stepping stones to our understanding of complicated phenomena.

The Orbiting Electron Model of the Atom

Every atom has a massive central nucleus, which carries a positive electric charge. Around the nucleus, and at a comparatively long distance away from it, revolve the electrons. These are minute, negatively charged particles which whirl around the nucleus much as the planets orbit around the sun. (Fig. 9.3.) Since oppositely charged objects experience a force of attraction, the positively charged nucleus attracts the negatively charged electron. This force maintains the electrons in their orbits and stops them from flying off at a tangent.

We cannot show on a diagram just how small the nucleus is compared to the electron cloud. Even a large nucleus has a volume which is less than a millionth of a millionth of the volume of the atom as a whole. The masses of nuclei vary from one atomic mass unit (1.66×10^{-27} kg) in the case of hydrogen to 238 u in the case of uranium. The mass of the electron (9.11×10^{-31} kg) is only

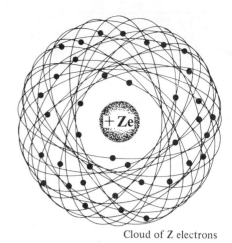

Cloud of Z electrons

Fig. 9.3 Atom of element no. Z (not to scale)

about one two-thousandth part of the mass of the hydrogen nucleus.

Each electron carries the same basic unit of charge (-1.6×10^{-19} C). The positive charge on the nucleus is always a whole number of these units. That is,

$$\text{Nuclear charge} = Z \times 1.6 \times 10^{-19} \text{ C}$$

The number Z is called the atomic number of the element.

The number of orbiting electrons equals the number of positive charges on the nucleus, thus preserving the electrical neutrality of the atom as a whole. For example, the hydrogen nucleus carries one positive charge and therefore has only one orbiting electron. (Fig. 9.4.) The helium atom has a doubly charged nucleus and two electrons and so on, each successive element having a progressively greater number until we reach uranium, which has 92 positive charges on the nucleus and 92 orbiting electrons. A complete list of the elements is given in Table 9.1.

Quantum Theory

Both the number and the arrangement of electrons outside the nucleus play a major role in the determining of the chemical and physical properties of the atom.

The electron orbits may be divided according to their size into a number of shells. Each shell has a limited number of vacancies for electrons. The first

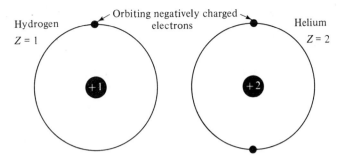

Fig. 9.4 The atoms of hydrogen and helium

shell can accommodate two electrons, the second shell can accommodate eight electrons. In general, the nth shell will hold a total of $2 \times n^2$ electrons.

The electrons required to match the nuclear charge first fill up the orbits nearest the nucleus of lowest energy, leaving the upper shells vacant.

The electronic structure of some of the simpler elements is shown diagrammatically in Fig. 9.5. Hydrogen is the simplest element and contains a single electron in the first shell. Helium, of atomic number 2, has two electrons which occupy and fill the first shell. The next element lithium has three electrons, one of which begins the second shell. Beryllium, boron, carbon, oxygen and fluorine progressively have one more electron in the second shell, until with neon the shell is full. The third shell begins with sodium and ends with Argon and so on. In all there are seven major shells which normally accommodate the electrons of the 100 or so elements.

Ionization Potential

The ionization potential of an element is the energy in electron-volts, required to remove the most loosely bound electron from the atom in its normal state. The electron-volt, the energy acquired by an electron in falling through a potential of 1 volt, is a convenient unit for expressing the energies of elementary particles.

$$1 \text{ eV} = 1.6 \times 10^{-19} \text{ J}$$

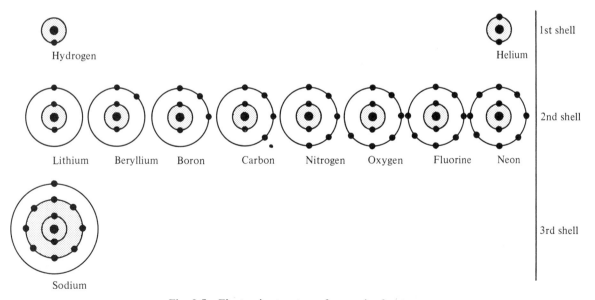

Fig. 9.5 Electronic structure of some simple atoms

The variation of ionization with atomic number is shown in Fig. 9.6. Notice how the ionization potential alters periodically with the atomic number.

Helium ($Z = 2$) and neon ($Z = 2+8$) have peak potentials because they have filled outer shells. They are followed by lithium and sodium, which have low ionization potentials, they have a single electron in the outermost occupied shell. Which is easy to remove because it is shielded from the nucleus by the completed inner shells.

Electrical Conductivity

Electrical conductivity depends on the existence of large numbers of mobile charge carriers. Substances such as sodium and aluminium, having only one or two electrons in the new shell, are good conductors and are classed as metals. The application of even a small electric potential causes their electrons to drift freely in the direction of the field.

There is a gradual change in the electrical properties as the shell becomes filled. An almost complete filled shell structure results in a material which is an insulator. Then at an intermediate stage there are materials which can be classed as neither metals nor insulators, the so-called semi-conducting materials such as carbon, silicon, and germanium.

Chemical Properties

The extent to which the outermost occupied shell of an element is complete is very important in determining its physical and chemical properties. When a main shell (or a particularly important subshell as explained later) is completely full, the element is very stable and rarely reacts with other elements. These inert substances are all gases at

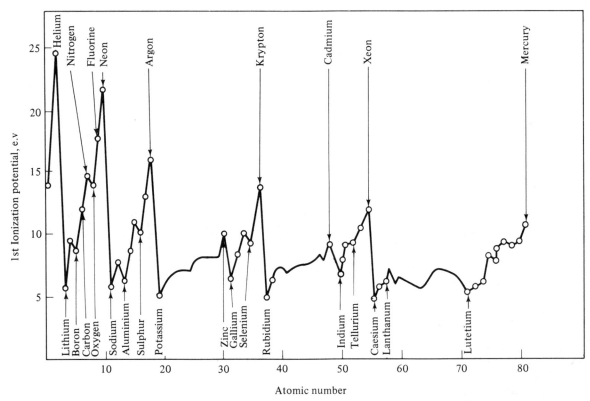

Fig. 9.6 Ionization potentials of the elements. (These potentials refer to isolated atoms and not to atoms grouped as in a molecule or a solid)

normal temperatures and pressures and comprise helium, neon, argon, krypton, xenon and radon. When atoms combine chemically they tend to do so in a way which imitates the structure of the inert gases, i.e., a filled outer shell.

Elements approaching the inert gas structure, exert a strong attraction for electrons, and even attract the electrons in the outer shell of another element. The annexing of extra electrons in this way leaves the receiving and donating atoms electrically charged. For example, in Fig. 9.7 the chlorine atom has attracted an extra electron from the sodium giving both atoms an inert

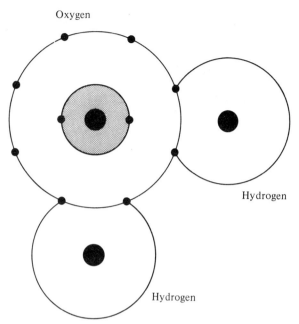

Fig. 9.8 By sharing electrons, each atom in the water molecule has achieved an inert gas structure

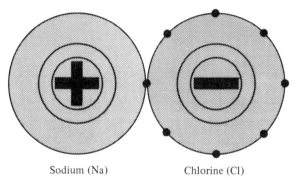

Sodium (Na)   Chlorine (Cl)

Fig. 9.7 An ionic bond between a sodium atom and a chlorine atom forms a molecule of sodium chloride

gas structure. In addition the chlorine has an excess of negative charge and the sodium an excess of positive charge, and the atoms become bonded together by electrostatic attraction. What has been described is the chemical reaction of sodium and chlorine, which results in the formation of sodium chloride, or common salt. Figure 9.8 shows another way in which the inert gas structure can be achieved by a sharing of external electrons between atoms.

Atoms whose outer shells have either just begun or are almost full of electrons are chemically highly reactive. This is the case for potassium, a metal which is so reactive that it ignites spontaneously when in contact with water. It will even react with ice and so potassium must be immersed in an inert liquid for safe storage. Chlorine, with an almost completed shell, is a corrosive gas. The most reactive substance known is fluorine which has so great an affinity for electrons that it extracts electrons from inert materials and even attacks glass.

The Periodic Table

If we arrange the elements in a table in order of atomic number or according to their ionization potentials, we can make useful generalizations about neighbouring elements. (Fig. 9.9.)

The inert gases helium, neon, argon, krypton, xenon and radon have peak ionization potentials and they are arranged in a column on the extreme right. The very stable structure occurs when the first and second shells are filled with their electrons.

The strongly metallic substances, lithium, sodium potassium, etc., have one electron in excess of the inert gas and have the lowest ionization potential, and they are placed in a column on the extreme left.

A group of similar reactive elements called halogens; fluorine, chlorine, bromine, iodine, and astatine are in the column before the inert gases. The semi-conductors carbon, silicon, and germanium are in group 4.

The minor peaks and troughs in the ionization potentials in Fig. 9.6 are explained by subdivisions

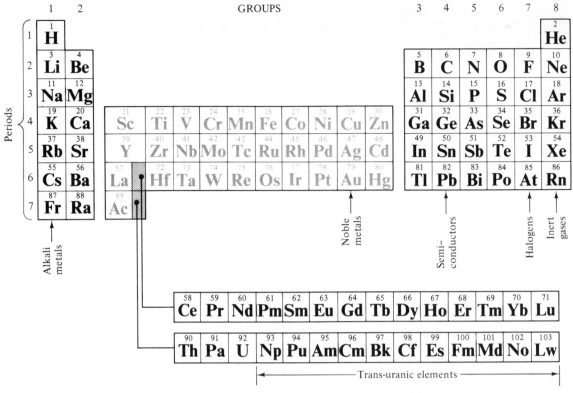

Fig. 9.9 The periodic table. Elements of similar properties are grouped in vertical columns

of the main shells into subshells. Students of chemistry will need to explore these subshells further. For the rest it is enough to understand the main features of the orbiting electron model, and the idea that the shell structure determines chemical and electrical properties.

Into the Nucleus

A dense positively charged nucleus is a feature of all atoms. Compared with the orbiting electrons, the nucleus of an atom exists in a serene world. The electron shells screen the nucleus from external influences so that it is not affected by physical stresses or electrical disturbances, and is quite unaffected by any chemical reaction involving the atom as a whole.

But what is the structure of the nucleus iteself? The nucleus contains two kinds of particle.

1. PROTONS of mass 1 u and charge +1 electronic charge
2. NEUTRONS of mass 1 u and charge = 0

The proton and neutron have almost the same mass and are both described as nucleons. (Fig. 9.10.) The number of protons in a nucleus determines the positive charge on the nucleus, and hence the atomic number Z of the element to which it belongs. (Fig. 9.11.)

Isotopes

The number of protons in the nucleus of an element identifies the element and it is therefore constant, but the number of neutrons may vary for different nuclei of the same element. These different forms are called isotopes of the element.

The number of naturally occurring isotopes of an element varies from one, for the element gold, to ten, for the element tin. The average number per element is about three. A particular isotope is referred to by writing the total number of nucleons and the atomic number before the symbol for the element. For example, the nucleus of the main isotope of lithium contains three protons and four

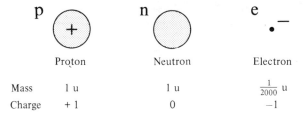

	Proton	Neutron	Electron
Mass	1 u	1 u	$\frac{1}{2000}$ u
Charge	+1	0	−1

Fig. 9.10 Comparisons of the approximate mass and charge on fundamental particles

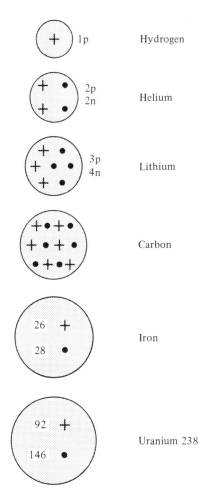

Fig. 9.11 Examples of the way protons and neutrons are combined in the nuclei of some atoms

neutrons and is represented by ^7_3Li. (Fig. 9.12.) It is also referred to as lithium 7. The other isotopes of lithium which can exist are written

$$\frac{\text{No. of nucleons}}{\text{nuclear charge}} \qquad ^5_3Li, \qquad ^6_3Li, \qquad ^8_3Li$$

As found in nature, one particular isotope of an element usually predominates, the other isotopes existing only in very small proportions, e.g., carbon consists of 98.89% of isotope $^{12}_6C$, the remainder being the isotopes $^{13}_6C$ and $^{14}_6C$. For this reason the atomic weights of elements are very near to whole numbers. Table 9.2.

Nuclear Stability and Radioactivity

Knowing, as we do, that like charges repel, how do we explain that protons in the nucleus in such close proximity do not burst apart? There must be even greater forces of attraction operating at close

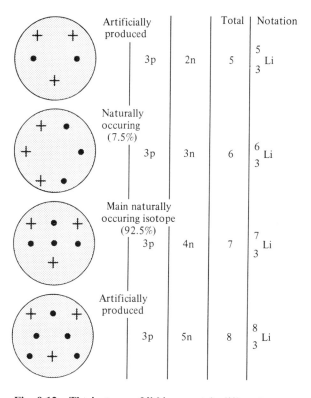

Fig. 9.12 The isotopes of lithium contain different numbers of neutrons. The one containing four neutrons is by far the most common

97

Table 9.2   Atomic Weights of the Isotopes of Some Elements

Symbol	Element	Nuclear charge (atomic number Z)	Relative atomic mass (u)	Mass number (A)	Abundance when naturally occurring
$^1_1$H	Hydrogen	1	1.00782	1	99.985%
$^2_1$H	Hydrogen (deuterium)	1	2.01410	2	0.015%
$^3_1$H	Hydrogen (tritium)	1	3.01605	3	
$^3_2$He	Helium	2	3.01603	3	0.00013%
$^4_2$He	Helium	2	4.00260	4	$\simeq 100\%$
$^5_2$He	Helium	2	5.01229	5	
$^{12}_6$C	Carbon	6	12.0000	12	98.892%
$^{13}_6$C	Carbon	6	13.0033	13	1.108%
^{234}U	Uranium	92	234.0397	234	0.006%
^{235}U	Uranium	92	235.0428	235	0.714%
^{238}U	Uranium	92	238.0496	238	99.280%

Although the system was conceived with the hydrogen atom [$^1_1$H] having unit mass it is more convenient to take the carbon 12 atom to have a mass of exactly 12.000 which gives the hydrogen atom a mass slightly different from 1.

quarters between both protons and neutrons. As the number of protons increases, so does the electrostatic repulsion and, therefore, to produce stability, more neutrons are needed, which add to the attraction but not to the repulsion. The lighter nuclei contain approximately equal numbers of neutrons and protons but as the size increases a greater proportion of neutrons is needed for stability. In the heaviest nuclei the neutrons outnumber the protons one and a half times.

The electrostatic repulsion in the nuclei of elements of more than 83 protons cannot permanently be counteracted by the short-range attractive forces and the nuclei are unstable. Some of the elements of atomic number greater than 83 exist in nature, while others can be produced artificially, but they are all radioactive and in the process of changing to elements of lower atomic number. This process we describe as radioactive decay or disintegration.

The three most common types of emission from radioactive isotopes are named after the first three letters of the Greek alphabet: alpha (α)-particles, beta (β)-particles, and gamma (γ)-radiation. (Fig. 9.13.)

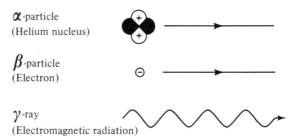

$\boldsymbol{\alpha}$-particle
(Helium nucleus)

$\boldsymbol{\beta}$-particle
(Electron)

γ-ray
(Electromagnetic radiation)

Fig. 9.13   The major types of radioactive emanation

Alpha-Particles

Alpha-particles have a mass of 4 u and carry a charge of $+2$ elementary charges. In other words they are helium nuclei consisting of two neutrons and two protons. When an atom emits an alpha-particle its mass number reduces by four, and its atomic number (positive charge) reduces by two, i.e., it becomes a different element. For example, radium decays by emitting an alpha-particle and forms radon.

$$^{226}_{88}\text{Ra} \rightarrow {}^{222}_{86}\text{Rn} + {}^4_2\alpha$$

The penetrating powers of the three components of the radiation vary widely. Alpha-particles with typical energies are absorbed by a few centimetres of air and they produce dense ionization of the air as they pass through it. Aluminium foil about 0.01 mm thick will eliminate them from radioactive emission.

Beta-Particles

Beta-particles are simply fast moving electrons. A neutron in the nucleus emits the electron and becomes a proton. As a result of beta decay a nucleus increases its positive charge by one and thus changes into the element one higher in atomic number.

$$\text{Thorium} \rightarrow \text{Protoactinium} + \text{Beta}$$
$$^{234}_{90}\text{Th} \rightarrow \, ^{234}_{91}\text{Pa} + \, _{-1}\beta$$

The most energetic beta-particles from some reactions will penetrate a centimetre of aluminium. Normally, beta-particles have energies far less than this and about 3 mm of aluminium acts as an effective shield.

Gamma-Radiation

Gamma-rays are electromagnetic rays of very short wave-length and they are generally the most penetrating of the radioactive emissions. The higher energy gamma-rays will penetrate up to a quarter of a metre of steel or a metre of concrete. (Fig. 9.14.) Gamma-radiation does not change the mass number or the atomic number of the emitting nucleus. It carries away the energy released by a re-arrangement of the nucleus. Since this re-arrangement often follows an alpha- or a beta-emission, sources of alpha- and beta-radiation commonly, but not always, emit gamma-rays.

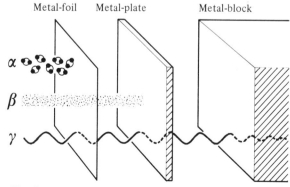

Fig. 9.14 Penetrating powers of radioactive emanation

Rate of Disintegration

The rate of disintegration of a radioactive nucleus is completely independent of its physical and chemical states. For example, one gram of radium emits 3×10^{-10} alpha-particles per second, whether it is at the centre of the earth at 4000°C and at a pressure of hundreds of times that of the atmosphere, or in a vacuum at the absolute zero of temperature. The rate of decay is directly proportional to the number N, of radioactive atoms present.

$$\text{Number/second} \propto \text{N}$$
$$\text{n/s} = \lambda \text{N}$$

where λ is a constant for a particular process and is called the decay constant.

Figure 9.15 shows how the number of radium atoms reduces with time. An important quantity which helps us to describe radioactivity is the half-life. The half-life is the time taken for half the number of atoms to disintegrate. Thus after a period equal to the half-life has elapsed only half the original number of atoms remain unchanged. The half-life T is simply related to the decay constant λ for any particular decay process.

$$T = \frac{0.693}{\lambda}$$

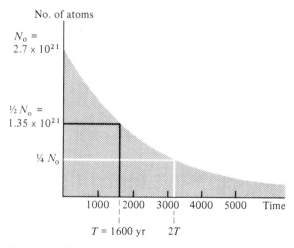

Fig. 9.15 The graph shows the way in which the number of atoms in one gram of radium reduces with time. An important constant is the half-life or period, T, which is the time taken for half the number of radium atoms to disintegrate

Radioactive Series

A nucleus may undergo a number of changes before it finally becomes stable. Table 9.3 shows the series of stages in the decay of uranium 238 into lead 206. All of these products are present in naturally occurring uranium. Two other major radioactive series exist, one starting with thorium 232 and the other with uranium 235 and both end in different isotopes of lead.

Table 9.3   The Uranium Series
Uranium 238 changes into lead 206 by emitting eight alpha-particles and six beta-particles. Gamma-radiation is also emitted as a result of the energy change on rearrangement of the nucleons.

Element	At. no.	Mass no.	Particle ejected	Approx. half-life
Uranium	92	238		
			α	4.5×10^9 yr
Thorium	90	234		
			β	25 days
Protoactinium	91	234		
			β	1.1 min
Uranium	92	234		
			α	300 000 yr
Thorium	90	230		
			α	83 000 yr
Radium	88	226		
			α	1600 yr
Radon	86	222		
			α	4 days
Polonium	84	218		
			α	3 min
Lead	82	214		
			β	27 min
Bismuth	83	214		
			β	20 min
Polonium	84	214		
			α	10^{-6} s
Lead	82	210		
			β	22 yr
Bismuth	83	210		
			β	5 days
Polonium	84	210		
			α	140 days
Lead	82	206		infinite

Detection of Radioactivity

The Geiger counter is one of the most rugged and versatile devices for detecting and counting any

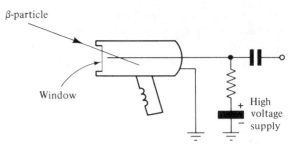

Fig. 9.16   A particle passing into the Geiger counter produces ionization which allows a pulse of charge to flow

particles or radiation which produce ionization. (Fig. 9.16.) A potential of several hundred volts is applied between the central wire electrode and an outer cylinder. The pressure of the gas in the cylinder and the applied voltage are such that an electrical discharge is on the point of occurring. The ionization caused by a particle generates a pulse of electricity, which is suitably amplified and fed to a counter or made to produce an audible click in a speaker.

A scintillation counter, another method of detection, uses the flash of light emitted by some crystals when they are struck by an alpha-, beta-, or gamma-ray. The light releases electrons which are multiplied in stages by repeated acceleration and collision in a photo-multiplier. (Fig. 9.17.)

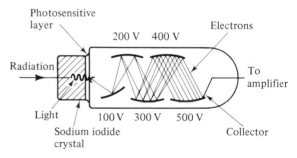

Fig. 9.17   A scintillation counter incorporates a photo-tube which multiplies the number of electrons in the pulse

When an ionizing particle passes through a space which is super-saturated with water vapour, the water condenses on the ions, and makes the path of the particle clearly visible. This is the principle of the cloud chamber, in which the paths of individual particles can be photographed and studied.

Uses of Radioactivity

When gamma-rays are absorbed by living matter they kill cells. This ability to destroy tissue can be used in the treatment of malignant growths in the body. The gamma-rays are kept directed at the tumour while the body is rotated round it. By this means, the malignant growth receives the largest exposure and a minimum of healthy tissue is destroyed.

Some elements which are not normally radioactive can be made artificially radioactive by subjecting them to bombardment by high energy particles from an atomic pile or a particle accelerator. These radioactive isotopes behave chemically exactly the same as a non-radioactive material, but their presence can be detected by their radioactive emission. Consider now how useful this detectable material can be. If a radioactive isotope of a material that can only be absorbed by a tumour, is injected into the blood stream, the exact limits of the growth can be determined by a radiation detector. The rate of assimilation of a substance into the body can be found by feeding a person a radioactive form of the substance and detecting its arrival at any particular part of the body. Many of these tracer techniques are used in the study of plant and animal processes.

The same property of radioactive radiation, which can affect the human embryo so disastrously can also produce changes in the reproductive cells of, for example, a particular strain of wheat. These mutations, as they are called, are mostly inferior to the form which has evolved naturally, but certain mutations may be useful and lead to a hardier or more productive type of wheat.

The intensity of a beam of gamma-rays depends very precisely on the thickness of a particular material which it traverses. We can use this property to measure and control the thickness of continuously produced material, such as metal plate. Gamma-rays can be used to detect internal cracks or flaws in metal castings in a similar way.

Radioactive Dating

During their growth, plants absorb carbon from the carbon dioxide in the atmosphere. A certain proportion of the carbon in the atmosphere is formed by the conversion of nitrogen into carbon by the cosmic rays. The carbon formed in this way is radioactive with a half-life of 5600 years. Thus, the tissue of plants and animals (since they eat the plants) is slightly radioactive, and one gram of the carbon from any living organism emits 16 beta-particles per minute. Once dead, the organic matter absorbs no more carbon 14, and as time passes the amount present decays until after 5600 years only eight beta-particles are emitted per minute per gram of carbon. (Fig. 9.18.) Thus, the

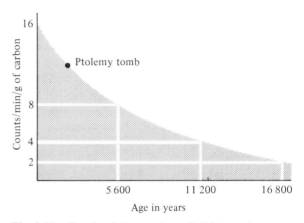

Fig. 9.18 Graph of the number of disintegrations per minute per gram of carbon, derived from organic sources, plotted against the time since it ceased to function

age of a fragment of bone or wood can be dated by measuring the activity and the carbon content of the specimen.

A similar technique may be applied to dating rocks containing uranium. With the help of such measurements, the age of the earth has been estimated at between 4 and 10 thousand million years.

Radiation Hazards

The ionization produced by radioactive emission is capable of killing living cells. Man has always been subject to ionizing radiation from the earth and to the cosmic radiation from space, and his existence shows that we can survive a certain amount of radiation. However, we must proceed with caution in changing the background level of radiation.

Radiative sources are described in terms of the

number of disintegrations which they undergo in unit time. The unit of activity is the becquerel (symbol Bq) defined as one disintegration per second.

Radiation, which passes unchanged through tissue, produces no damage. The damage is caused when the radiation is absorbed by the living cell. The unit of exposure to X-radiation or gamma-radiation is the Roentgen (symbol R) the radiation which will produce 2.08×10^9 ion pairs in a cubic centimetre of dry air at S.T.P. (i.e., 1.293×10^{-3} g of air).

$$1 \text{ Roentgen} = 1 \text{ R} = 1000 \text{ mR, etc.}$$

The radiation of any type actually absorbed, referred to as the dose, is measured in rads. The rad is the dose which imparts 10^{-5} joule of energy per gram of matter.

Normal background levels of radiation are such that in a year each person receives about 0.1 rad. There is no evidence that this amount of radiation is harmless, but it is one of the natural hazards that the human race is subject to, and very little can be done about it. It may be that this background radiation is what causes mutations and allows natural selection to operate.

If received over a short period, an exposure to 100 rad would make a person seriously ill and a dose of about 400 rad would be fatal to 50% of persons exposed. The question is where to set the limit to exposure compatible with human health. The effects of radiation may be very long term and they may not be easy to diagnose. For example, 20 years after the bombing of Hiroshima and Nagasaki hundreds of people still died as a result of the radiation received at that time. They may have died prematurely of conventional diseases because their resistance to disease has been undermined by the radiation. Even a small increase in the background is thought to increase the number of malformed babies born, and the incidence of certain diseases such as anaemia and cancer. As a result of our increasing understanding of the effects of radiation, our estimate of the amount of radiation that the human body can tolerate has been frequently revised, and with each revision the estimate has been reduced. To warn people of the presence of radioactive materials a sign has been internationally adopted. (Fig. 9.19.)

Fig. 9.19   Radioactive materials warning sign

Energy and Mass

Large energies are involved when nuclear particles combine to form a larger nucleus or split to form a simpler nucleus. The energy shows itself as a change in mass according to Einstein's equation:

$$E = mc^2$$

i.e.,

$$\text{Energy} = \text{mass} \times (\text{velocity of light})^2$$
$$\text{Joules} = \text{kg} \times (3 \times 10^8)^2$$

Because of the large factor (9×10^{16}) a small change of mass gives rise to a very large release of energy.

Fusion

When hydrogen nuclei collide at high speed they combine to form a helium nucleus. (Fig. 9.20.) The helium weighs less than the colliding hydrogen nuclei. The difference amounts to 4.6×10^{-29} kg.

$$\begin{aligned} \text{Energy } E &= mc^2 \\ &= 4.6 \times 10^{-29}(3 \times 10^8)^2 \text{ J/helium atom} \\ &= 62 \times 10^{16} \text{ J/kg of helium produced} \end{aligned}$$

This process of fusion is called a thermonuclear reaction because, to have the necessary velocity of collision, the hydrogen must have a temperature of several million degrees. Such temperatures exist in

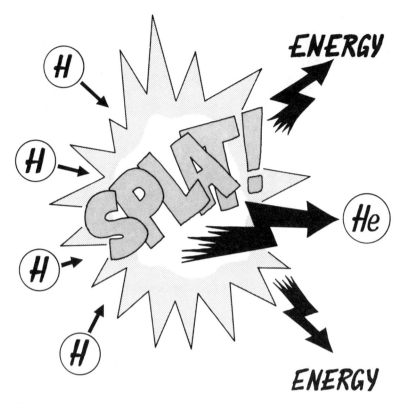

Fig. 9.20 Fusion—four hydrogen nuclei collide to give helium and energy

the interior of stars which maintain their energy output by the conversion of hydrogen to helium. The sun, a typical star, provides 3.3×10^{26} joules/s by this process. In consequence, the mass of the sun reduces by 4.2×10^9 kg/s, but, have no fear, this amounts to only one millionth part of its total mass in 10 million years.

Fission

Fission is the splitting of a nucleus into two or more fragments. A slow neutron colliding with a nucleus of uranium 235 may cause it to split into smaller nuclei of krypton and barium, releasing two extra neutrons. (Fig. 9.21.)

The total mass of the particles produced is less than that of the uranium atom and the neutron. During the reaction, this mass has become associated with other forms of energy such as kinetic energy of the product particles and electromagnetic radiation. The process releases millions of times more energy than would the combustion of a similar mass of material.

Note that the process of fission increases the number of free neutrons, some of which produce the fission of other uranium nuclei. Self-maintaining processes of this kind are called chain reactions. If every neutron released caused a further reaction, there would be a very rapid spread

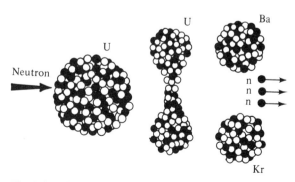

Fig. 9.21 One possible mode of splitting of uranium produces barium and krypton and three neutrons

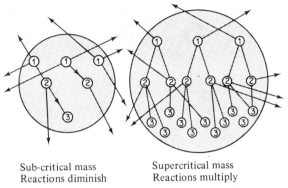

Sub-critical mass
Reactions diminish

Supercritical mass
Reactions multiply

Fig. 9.22 Schematic representation of three stages in a chain reaction in uranium 235

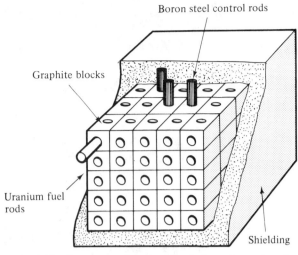

Boron steel control rods

Graphite blocks

Uranium fuel rods

Shielding

Fig. 9.23 In an atomic pile, the fission of uranium is controlled by retractable boron rods which absorb neutrons and slow down the chain reaction

of fission to all the uranium nuclei present. Not all the neutrons released succeed in producing fission. Most of them pass out through the boundary of the solid without making further collisions. In a large solid this escape is less likely and more of the neutrons produce fission. There is therefore a critical size above which the number of nuclei splitting grows in number. When the critical mass is only slightly exceeded, the material gets hotter and hotter; when the critical mass is grossly exceeded, it explodes. (Fig. 9.22.)

Atomic Pile

In an atomic reactor a mass of uranium is kept in a near critical condition by interspersing it with other materials. (Fig. 9.23.) Carbon slows down the product neutrons to thermal velocities which makes them more likely to initiate further fission. The boron rods on the other hand absorb neutrons and these reduce the rate of fission. Thus, by changing the position of the boron rods, engineers control the rate of heat output.

A flow of liquid or gas over the reactor carries away the heat to a heat exchanger where it produces steam. (Fig. 9.24.) This steam drives a tur-

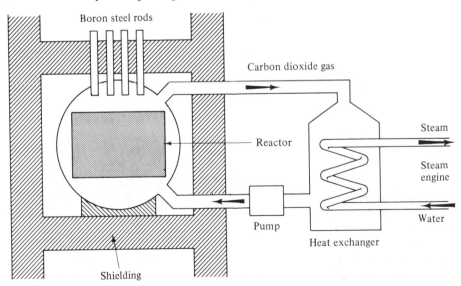

Boron steel rods

Carbon dioxide gas

Reactor

Steam

Steam engine

Water

Pump

Heat exchanger

Shielding

Fig. 9.24 Schematic diagram of one type of British nuclear reactor

bine which either delivers the power on the spot, or is used to generate electrical energy for distribution.

Nuclear Explosives

In an atomic bomb, several sub-critical masses of uranium 235 are brought together to form one super-critical mass. (Fig. 9.25.) When the combined masses reach the critical quantity, the uranium explodes spontaneously in a fraction of a second. The explosion, in fact, begins as the masses approach each other, so that, in order to give as much time as possible for fission to occur, the separate masses of uranium 235 are shot together by charges of conventional explosive.

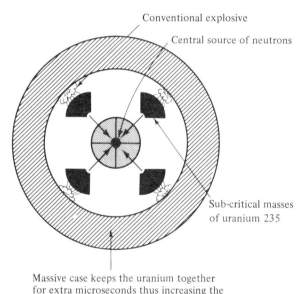

Conventional explosive

Central source of neutrons

Sub-critical masses of uranium 235

Massive case keeps the uranium together for extra microseconds thus increasing the efficiency

Fig. 9.25 A nuclear explosion is produced when several masses of uranium 235 are combined to produce a super-critical mass

Much bigger explosions can be produced by the fusion of hydrogen nuclei. In this type of reaction, a temperature of several million degrees is produced by means of a fissional explosion, and, at this enormous temperature, the thermal velocities of atoms are sufficient to force hydrogen nuclei to combine to form atoms of helium. In fact, only heavy hydrogen isotopes (deuterium and tritium) have been made to produce explosions.

The heat from repeated underground nuclear explosions may prove, in the future, to be an effective means of deriving nuclear power, if a way can be found to extract this heat through remotely positioned heat exchangers. Attempts to control and harness thermonuclear energy have so far proved unsuccessful. The potentialities of nuclear explosions as weapons of war are obvious and terrible, as the bombings of Nagasaki and Hiroshima demonstrated.

Questions

1. Why will we never be able to see the structure of an atom?

2. By what means may electrons be removed from atoms?

3. Why is the half life of radioactive materials given rather than the full life?

4. Why are free neutrons more difficult to detect than electrons or protons?

5. Which should man fear more about nuclear explosion, blast or pollution?

Multiple Choice

1. Isotopes of an element have the same
    (a) chemical properties, (b) appearance,
    (c) number of protons, (d) number of neutrons.

2. The most penetrating of the following radiations in air is
    (a) α, (b) β, (c) γ, (d) neutrons.

3. In an atom the nucleus contains
    (a) positive charge, (b) negative charge,
    (c) neutrons, (d) protons.

4. If a radioactive isotope has a half life of 4 hours the amount remaining after 12 hours will be
    (a) $\frac{1}{4}$, (b) $\frac{1}{8}$, (c) $\frac{1}{16}$, (d) $\frac{1}{32}$.

5. The element having the lightest atom is
    (a) oxygen, (b) nitrogen, (c) helium, (d) hydrogen.

6. β particles are
    (a) positively charged, (b) electromagnetic rays,
    (c) electrons, (d) negatively charged.

7. How many atoms are there in a diatomic molecule?
    (a) 1, (b) 2, (c) 3, (d) 4.

8. What is the ratio of masses of the hydrogen and helium atom?
    (a) 1:2, (b) 1:3, (c) 1:4, (d) 1:8.

9. The electron volt is a unit of
    (a) voltage, (b) energy, (c) charge, (d) power.

10. Atomic reactors use the process of nuclear (a) fusion, (b) combustion, (c) decay, (d) fission.

11. The removal of an electron from an atom is an example of (a) ionization, (b) fission, (c) fusion, (d) polymerization.

12. Einstein's relation between mass m and energy E is (a) $E = mc$, (b) $E = m/c$, (c) $E = m^2c$, (d) $E = mc^2$.

Multiple Choice Answers

1(abc), **2**(c), **3**(acd), **4**(b), **5**(d), **6**(cd), **7**(b), **8**(c), **9**(b), **10**(d), **11**(a) **12**(d).

10.
The Structure of Materials

Having recognized atoms and molecules as the basic building blocks from which matter is constructed, we turn our attention to the different arrangements of the molecules and the binding forces which hold them together.

An important feature of a substance is the state in which it exists, i.e., whether it is a solid, a liquid, or a gas. The state of a substance depends on the conditions of pressure and temperature which prevail. For example, an increase in temperature will change any solid into a liquid and ultimately into a gas or vapour. Substances tend to be classified according to their state under normal conditions. We class water as a liquid, but an Eskimo might not agree. The molecules in each state influence each other to different extents. We will describe the structure of solids first, and then examine the transition to other states later.

Solids

In a solid the molecules hold each other into a regular three-dimensional pattern. The mutual attraction of molecules increases as they approach each other, but it turns to strong repulsion, if they get very near. (Fig. 10.1.) At a certain distance apart the molecules are in equilibrium. The forces holding the molecules in position form links or bonds which can be stretched or compressed rather like springs. (Fig. 10.2.)

When an overall compression or stretching is applied to a body, the way it responds depends on its structure and on the strength of the links between molecules. The links can be very strong as is demonstrated by a 40 mm diameter steel cable which supports 50 tons quite safely.

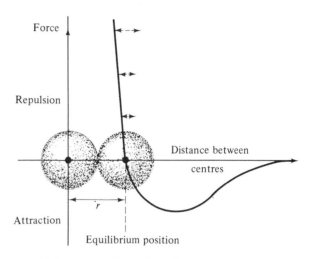

Fig. 10.1 **The resultant force between atoms changes from being attractive at long range to being strongly repulsive at short range. The amplitude of vibration and the average distance between atoms increases with temperature**

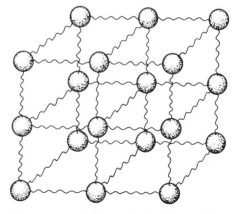

Fig. 10.2 **In a crystalline solid, the links hold the atoms into a regular pattern**

Crystal Structure

For the sake of convenience, molecules and atoms are often represented by spheres. In fact, they are not spherical but have definite shapes, and a tendency to link together in characteristic ways. The molecules of any particular substance have a set number of links, which reach out from the molecule in certain directions. For these reasons most solids are crystalline, that is, they are made up of a geometrical pattern of atoms. (Fig. 10.3.) The crystal structure of a solid can often be deduced from the shape of its crystals (Fig. 10.4) or by a microscopic examination of a fractured surface. (Fig. 10.5.) Crystal structure is analyzed by X-rays which penetrate the metal and are strongly reflected by the crystal planes. At one time, only the more obvious and simply shaped solids were thought to be crystalline but more rigorous examination shows that almost all solids are crystalline. (Fig. 10.6.)

The particular crystalline pattern is a very important factor in determining the properties of a solid. The difference between two arrangements of the same atoms can be very pronounced. Graphite and diamond are solids having two different arrangements of the carbon atom. One is an excellent lubricant and the other a superlative abrasive. (Fig. 10.7.) The carbon atom has four links. In graphite three of these links lie in a plane joining up with other atoms to form a sheet structure. The sheets are only loosely joined together by the fourth link of the atoms and it is very easy for the sheets to slide over each other. In diamond the four links are arranged symmetrically around each atom, forming a stable pyramid structure.

Metals

Metals as a group of substances make a vivid impact on us because they share special properties which we can readily recognize. They shine, i.e., have a metallic lustre. They are good conductors of heat and electricity. They have high tensile strengths and they can be pressed into shapes or pulled out into wires without breaking.

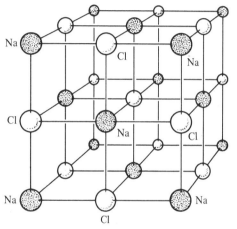

Fig. 10.3   Sodium and chlorine ions fit together forming a cubic structure

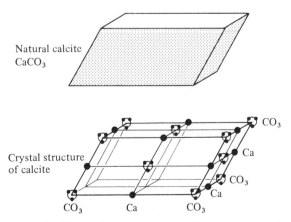

Fig. 10.4   The shape of the crystal indicates the crystal structure

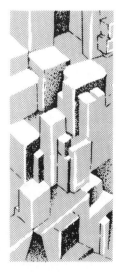

Fig. 10.5   Fractured surface of aluminium indicates a cubic crystalline structure

Fig. 10.6 Naturally occurring quartz crystals
(*Courtesy the Institute of Geological Sciences*)

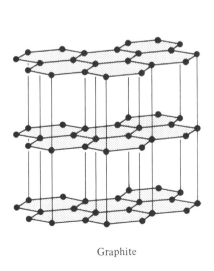

Graphite

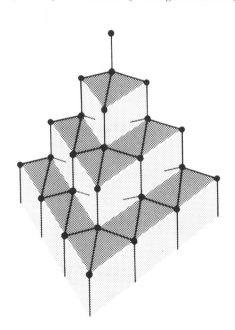

Diamond

Fig. 10.7 Two crystalline forms of carbon

Metals contain a large number of highly mobile negatively charged electrons which can move freely through the metal's crystalline structure. These electrons carry electric current easily, they pass on thermal vibrations quickly, and at the surface they oscillate in unison with light waves giving efficient reflection.

Extraction of Metals

Metals apart from gold are not usually found in nature in the metallic state. They are found as ores which may be chemical combinations of the metal with oxygen or sulphur. The ore is mined, crushed and separated from the dirt and unwanted matter. Next it is converted into crude impure metal by the process of smelting which breaks up the chemical combination. Finally the metal is refined to the pure metal by a blast of air to oxidise the impurities or by electrolysis. It is cast into convenient shapes, pigs, ingots, slabs or bars for future treatment. It may then be rolled into sheets, forged into shapes or drawn through dies to make wire.

Crystalline Metal Structures

It is the crystalline structure of metals which accounts for many of their properties. (Fig. 10.8.) The bonds between the atoms in metals are strong but flexible. They move and they stretch but they do not break. The crystal planes slip one over the other allowing a change of shape without fracture. As a metal solidifies the first minute crystals form in the liquid. These grow to form extended crystals called dendrites. They continue to grow until they reach the region occupied by another dendrite. Now the region near each dendrite forms as a complete crystal which we call a grain. (Fig. 10.9.)

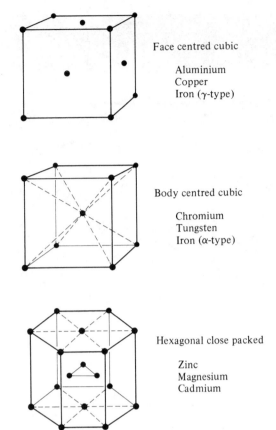

Face centred cubic

Aluminium
Copper
Iron (γ-type)

Body centred cubic

Chromium
Tungsten
Iron (α-type)

Hexagonal close packed

Zinc
Magnesium
Cadmium

Fig. 10.8 Metallic crystal structures

The grain structure of metals plays a big part in determining their physical properties. In some cases the grain size is microscopic but in others it is easily seen as in the case of newly galvanized surfaces.

A metal which can be cold drawn through dies into wire is said to be ductile. A metal which can be easily rolled or hammered into extended shapes

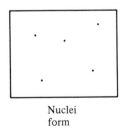

Nuclei
form

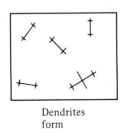

Dendrites
form

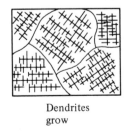

Dendrites
grow

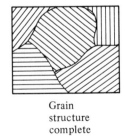

Grain
structure
complete

Fig. 10.9 Dendritic growth. The grain structure arises from the extension of dendrites as solidification progresses

110

is said to be malleable. (Fig. 10.10.) These processes gradually reduce the size of the grains which causes the material to harden and become brittle, less ductile and less malleable.

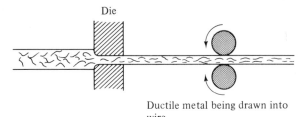

Ductile metal being drawn into wire

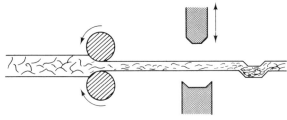

Malleable metal pressure rolled into sheet or formed into shapes

Fig. 10.10   Reduction of grain size in cold working

Crystal Defects

Although many crystals appear quite symmetrical, absolutely perfect crystals are extremely rare. Most crystals possess defects which are so small that they may not be visible even with the most powerful microscopes, but which nevertheless exert a very strong influence on their properties. (Fig. 10.11.) Thermal vibration causes the defects to diffuse at random about the crystals. When under stress the defects tend to accumulate and cause deformation of the crystal under much less force than that which would deform a perfect crystal. (Fig. 10.12.) To make a material remain elastic under greater stress, we must either reduce the number of dislocations or stop their movement.

Fatigue

Materials which are subject to repeated stresses over a period may break or become deformed under loads well below the breaking strain. This phenomenon is called fatigue and is caused by the accumulation in use of dislocations in one particular plane or direction. (Fig. 10.13.) Centres for the concentration of defects may be created by minor damage during manufacture or assembly, e.g., a chance hammer blow.

In the design of structures, such as bridges and machines, a very large margin of safety must be allowed between the maximum load to be applied in use and the breaking strength of the structure. The ratio of breaking strength to maximum load, called the safety factor, is usually about 4 for steel structures and about 10 for brick structures. Fatigue failure occurs less where high safety factors apply, but the risk can also be lessened by good design. (Fig. 10.14.) This type of failure could be disastrous in an aeroplane. Thus components subject to fatigue are replaced after a certain period of use, even though they may show no superficial signs of wear or deterioration.

Polycrystalline Materials

The usual methods of hardening materials aim at immobilizing the dislocations. Most materials are polycrystalline, that is, they are made up of a large number of crystals joined at the boundaries. (Fig. 10.15.) These boundaries restrict the movement of the dislocations and give strength to the material. The greater the number of boundaries the stronger the material becomes and vice versa. If we wish to harden a material, we cool it rapidly and it crystallizes in many places at once. To make it softer, we cool it slowly after heating (annealing) and larger crystals form. (Fig. 10.16.) Metals are sometimes hardened by cold·working, that is, by being hammered or rolled into sheets, which increases the area of the boundaries between crystal zones and restricts the movement of dislocations. The prestressing of the material allows the more mobile defects to move and, perhaps, become fixed before the material is machined to shape.

Whiskers

Until whiskers were discovered, scientists thought it would be impossible ever to produce crystals which were free from the effects of dislocations. Whiskers are minute metal filaments, which grow on the metal coatings of electronic components.

During experiments aimed at eliminating these filaments, it was found that the whiskers in tin were about 1000 times as strong as tin wire of the same diameter. We attribute this strength to the fact that each whisker is a single dislocation-free crystal. Unfortunately, they are so minute, only about 10^{-3} m in diameter, that we cannot as yet make use of their strength. Further development, however, may one day produce a material of unprecedented toughness from these whiskers.

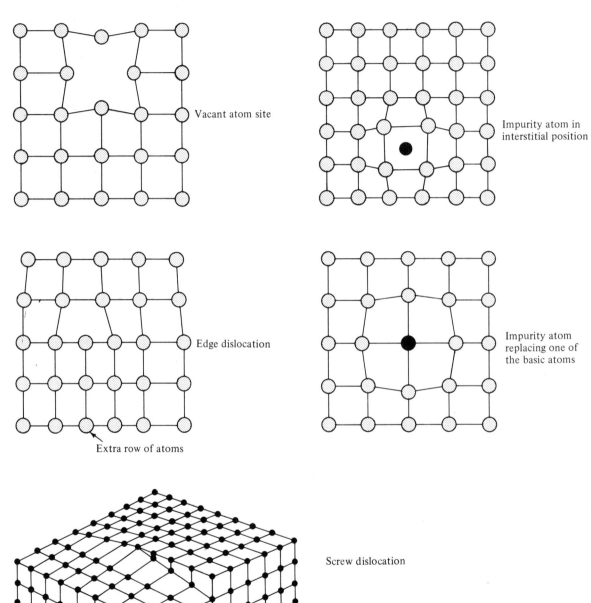

Vacant atom site

Impurity atom in interstitial position

Edge dislocation

Impurity atom replacing one of the basic atoms

Extra row of atoms

Screw dislocation

Fig. 10.11   Crystal defects

Alloys

An alloy is composed of a metal mixed with other metals or impurities. The introduction of impurities into a crystalline material can change its properties drastically. For example brass is harder than either copper or zinc which are its constituents. Constantin has ten times the electrical resistance of either copper or nickel of which it is composed in the ratio of 6:4. The impurity makes the lattice less regular and prevents crystal deformation along slip planes and also impairs the flow of electrons.

An alloy is usually made by mixing the constituents in the molten state and then allowing the mix to cool. In some cases the components crystallize out separately and the alloy is a composite of crystals of two metals. Plumbers solder is a mixture of tin and lead crystals and is an example of this type of alloy. Such alloys are referred to as eutectics.

In other types of alloy the components are completely dispersed, i.e., dissolved in each other. On solidification the alloy is homogeneous and is described as a solid solution. Copper and nickel form solid solutions in each other at all temperatures. Other solid solutions are brass (copper zinc) and bronze (copper tin).

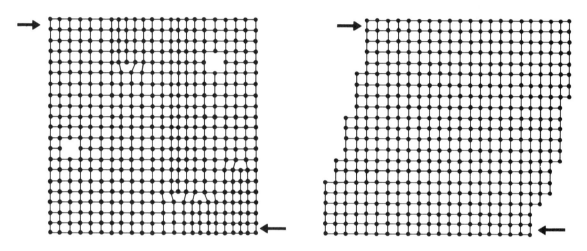

Dislocations move and material permanently changes its shape

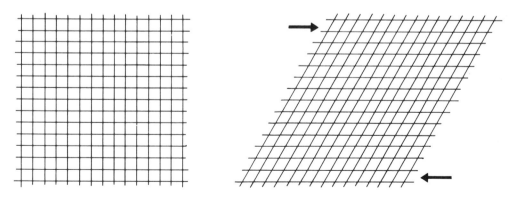

Perfect crystal recovers its shape

Fig. 10.12   Effect of force on a perfect crystal and one with dislocation

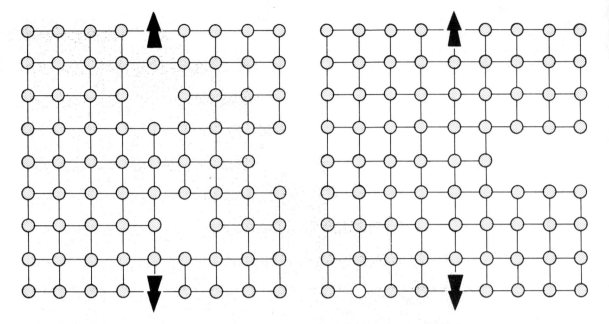

Fig. 10.13 Surface defect acts as a point at which other defects collect and develops into a crack

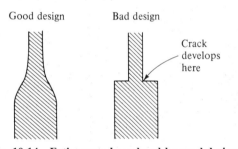

Fig. 10.14 Fatigue can be reduced by good design

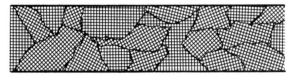

Fig. 10.15 Most materials are polycrystalline and are made up of crystallites of different lattice orientation

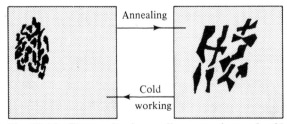

Fig. 10.16 The result of annealing a specimen of cold worked brass is an increase in the size of the crystal which makes the brass less brittle

Steel

Iron in its pure state is almost useless because of its softness. Minute amounts of carbon of say 0.25% as in mild steel makes it much stronger. High carbon steel has about 1% carbon and is hard enough to use as cutting tools. Cast iron used for engine blocks has about 3% carbon and smaller amounts of silicon and manganese. Stainless steel contains a large proportion (13–18%) of chromium as well as traces of carbon and manganese. It is highly resistant to corrosion and has little or no magnetic influence.

The final properties of steel depend very strongly on the heat treatment and the physical treatment it has received. Rapid or gradual cooling from the melt will result in different crystalline structures and grain size giving harder or softer properties. (Fig. 10.17.)

Welding, Soldering, Brazing

We can join metals by riveting, bolting and glueing but in some cases a better joint can be made by welding, brazing or soldering them together. In

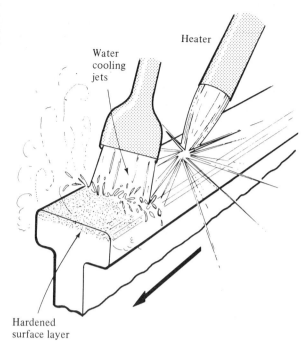

Fig. 10.17 **Surface hardening steel rail by a continuous process of heating and water quenching**

Heater

Water cooling jets

Hardened surface layer

welding the surfaces are heated until they melt and fuse one to the other. This requires a temperature above the melting point of the material to be welded. Solder is an alloy which melts and fuses with the two surfaces to be joined. Tinman's solder of lead 40% and tin 60% melts at about 200°C. Brazing involves the melting of brass to join the two surfaces. It is stronger than solder but it requires a higher temperature of about 850°C.

Plastics

Plastics are so called because under certain conditions they can be moulded plastically. We classify them into three main types.

1. Thermoplastics are soft when heated and become rigid when cooled.
2. Thermosetting plastics undergo a chemical change when heated and become rigid. They do not melt when heated further.
3. Cold setting plastics are usually made by mixing liquids which react chemically and then become rigid.

Rubber was the first plastic and was made from the sap of rubber trees. Other plastics are made from animal, vegetable and mineral sources. All the plastics have one other thing in common; they are made up of very long chain molecules which contain the carbon atom. The formation of these long molecules from smaller molecules is polymerization and the end product is called a polymer. Examples of polymers, polyethylene (polythene) and polyvinylchloride (P.V.C.) are given in Fig. 10.18. Plastic products include nylon rope, terylene fabrics, and polystyrene foam, which is formed when steam meets beads of polystyrene. (Table 10.1.)

Plastics are light materials relatively cheap to produce and easily coloured. They are good electrical and thermal insulators due to the absence of free electrons for transmission. They are highly resistant to solvents and to chemical attack. They lend themselves to mass production by compression-moulding techniques.

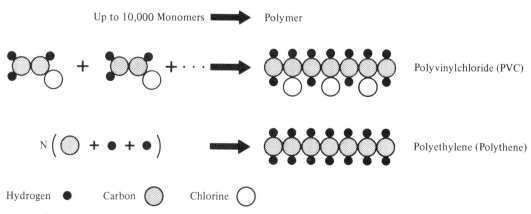

Up to 10,000 Monomers ➡ Polymer

Polyvinylchloride (PVC)

Polyethylene (Polythene)

Hydrogen ●   Carbon   Chlorine ○

Fig. 10.18 **Formation of PVC by polymerization. The chains may contain tens of thousands of molecules**

Table 10.1 Plastics

General name	Other names	Use
Vinyls	Polyvinyl chloride	Toys, garments, table covering, pipes
	Polyvinyl acetate polythene	Gramophone records, kitchen equipment, wrappings
	Polypropylene	Containers resistant to hot water
Polyamides	Nylon	Clothes, ropes bearings
Polyester	Terylene (dacron)	Clothes, ropes, sails
Acrylics	Perspex	Windows, displays
Phenobics	Bakelite urea formaldehyde	Electrical equipment
Polyurethanes		Foams, sponges, paint

Change of State

The linkages between the atoms or molecules of a solid can only restrain the atoms to a limited extent. Thus, if the thermal vibrations exceed this limit, the interatomic attractive forces are overcome and the atoms are less rigidly held in position. This is the case when a solid reaches its melting point.

The boiling or melting point can be obtained by plotting the graph of Fig. 10.19.

Crystalline substances which are pure, or which contain a uniform distribution of impurities, produce well defined horizontal sections. Non-crystalline or amorphous substances such as glass, paraffin wax, and many kinds of 'plastic' material do not have a well-defined melting point. They become softer or more plastic as the temperature rises. Even at low temperatures they flow and they may be regarded as liquids of very high viscosity. Glass windows taken from very old houses are thicker at the bottom than at the top due to the pull of gravity.

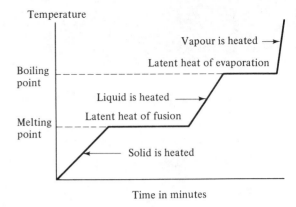

Fig. 10.19 **When a substance is supplied with heat at a constant rate, the temperature rise is delayed at the melting and boiling points, due to the absorption of latent heat**

Effect of Pressure on Melting Points

Most substances contract when they solidify, so that an increase in pressure assists the process of solidification, making it occur above the normal melting point. A few substances, including cast iron and water, expand on freezing, and an increased pressure lowers their melting points. The depression of the melting point of water by 1°C requires a pressure of about 10^7 N/m². Pressures of this order are produced under sledge runners and ice-skate blades. The ice melts momentarily and the runner slides on a film of water.

The melting point of a substance is usually lowered by the presence of an impurity. Sea-water freezes well below 0°C, and most alloys of two metals melt at a lower temperature than that of the major constituent.

Questions

1. Why do engineers avoid sharp angles and small radii when designing components?

2. What characteristics of an atom and molecule determine the shape of its crystalline form?

3. Why is a metal heated before forging?

4. Under what conditions would you form a crystal to keep it as free from dislocations as possible?

5. What makes rubber rubbery?

Multiple Choice

1. Which of the following is the most ductile?
   (a) cast iron, (b) copper, (c) carbon, (d) diamond.

2. The application of high pressure could cause ice to
   (a) liquify, (b) solidify, (c) evaporate, (d) crystallize.

3. Brazing is a joining process involving
   (a) solder, (b) brass, (c) glue, (d) welding.

4. A material which is softened by heat is described as
   (a) thermosetting, (b) thermoplastic, (c) crystalline,
   (d) amorphous.

5. A metal which can be hammered into extended shapes is said to be
   (a) brittle, (b) plastic, (c) malleable, (d) resilient.

6. Which material is not classed as a plastic?
   (a) nylon, (b) perspex, (c) graphite, (d) polythene.

7. All materials classed as plastics contain
   (a) polymers, (b) rubber, (c) carbon, (d) polythene.

8. Which processes are used to harden metals?
   (a) annealing, (b) quenching, (c) cold working,
   (d) melting.

9. Which processes are used in metal extraction and refining?
   (a) smelting, (b) galvanizing, (c) forging,
   (d) electrolysis.

10. A solid solution of two metals is called
    (a) an alloy, (b) a mixture, (c) an eutectic,
    (d) a solute.

Multiple Choice Answers

1(b), 2(a), 3(b), 4(b), 5(c), 6(c), 7(ac), 8(bc), 9(ad), 10(a).

11.
Liquids

Fluids is a general name for substances that flow. So liquids and gases both qualify as fluids. Here we shall look at liquids and start by summarizing some properties of liquids that spring partly from one common experience and partly from simple experiments. (Fig. 11.1.)

Under the force of gravity a liquid takes the shape of its container and it assumes a horizontal surface, i.e., it finds its own level. Below the surface a pressure is exerted which increases with depth, and does not depend on the shape of the container. Inside the liquid the pressure is exerted equally in all directions. Where the liquid is in contact with the container the pressure acts at right angles to the container.

Liquids

We explained some of the properties of solids in terms of the forces between the molecules but can these forces explain the properties of liquids? The cohesion is different in solids and liquids as you can feel by comparing the cutting of a solid with the drawing of a knife through a liquid. However, forces do exist between the molecules in liquids as we can see from the way in which liquids hold together. For example, the surface tension in a droplet is a result of such forces making the surface behave like an elastic film under constant tension.

The forces between the molecules of a liquid are comparable to those in the solid but they are continually being broken up by thermal vibrations.

The crystalline pattern disappears and molecules move through the extent of the liquid under the influence of even the smallest force.

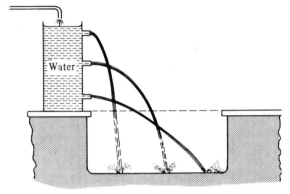

Pressure increases with depth

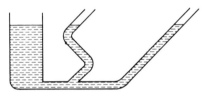

Liquid takes the shape of the vessel and comes to the sam horizontal level

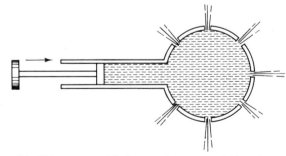

Liquid in contact with the container acts at right angles

Fig. 11.1 Experiments which show liquid properties

Pressure in a Liquid

Pressure is the force exerted per unit area

$$\text{Pressure} = \frac{\text{force}}{\text{area}}$$

The units are newtons per square metre which are given the special name of pascals (symbol Pa).

The pressure at a depth in a liquid depends on the depth h, the density ρ and the force of gravity. Figure 11.2 shows a column of liquid of cross-sectional area A and height h, and uniform density ρ

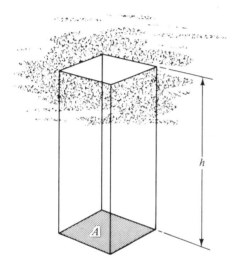

Fig. 11.2   The pressure at a depth in a liquid is due to the weight of the liquid

Volume of column $= hA$

Mass of column $=$ volume $\times$ density
$$= hA\rho$$

Force exerted by the column (weight)
$$= hA\rho g$$

where g is the acceleration due to gravity

$$\text{Pressure} = \frac{\text{force}}{\text{area}} = \frac{hA\rho g}{A}$$

$$\text{Pressure} = h\rho g$$

This gives the pressure in pascals (N/m^2) when h is in metres, ρ is in kg/m^3 and g is m/s^2.

Problem

What is the pressure at a depth of 100 mm in mercury of density 13.6×10^3 kg/m^3 assuming the gravitational acceleration is 9.8 m/s^2.

Solution

$$\text{Pressure} = h\rho g = 0.1 \times 13.6 \times 10^3 \times 9.8$$
$$= 13.3 \times 10^3 \text{ Pa}$$

Instead of expressing pressure as a force per unit area we can express it in terms of the height of a column of a reference liquid. For example the length of a column of water is used to specify pressure in the gas industry where they are dealing with small pressures in excess of atmospheric pressure. (Fig. 11.3.)

Thus 80 mm of water
$$= h\rho g$$
$$= 0.08 \times 1000 \times 9.8$$
$$= 784 \text{ pascals above atmospheric}$$

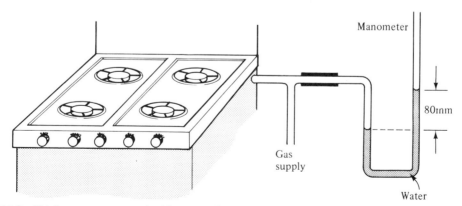

Fig. 11.3   U-tube manometer used with open end to record pressures in excess of atmospheric pressure

The height of a mercury column is often used to specify pressures since atmospheric pressure is measured by a mercury barometer and expressed in millimetres of mercury.

$$1 \text{ mm of mercury} = h\rho g = 0.001 \times 13.6 \times 10^3 \times 9.8$$
$$= 133.3 \text{ Pa}$$

Standard atmospheric pressure
$$= 760 \text{ mm mercury}$$
$$= 760 \times 133.3 = 1.013 \times 10^5 \text{ Pa}$$
$$\simeq 10^5 \text{ Pa}$$

For larger pressures the atmosphere itself may be used as a rough unit, e.g., a diver's air cylinder may contain air at a pressure of 10 atmospheres.

Divers also use a pressure gauge to measure their depth. The pressure increases by 1 atmosphere every 9 m of descent.

Table 11.1 Units of Pressure

Unit	Symbol	Value
Pascal	1 Pa	$= 1 \text{ N/m}^2$
Bar	1 bar	$= 10^5 \text{ N/m}^2 = 10^5 \text{ Pa}$
Millibar	1 mb	$= 10^2 \text{ N/m}^2 = 10^2 \text{ Pa}$
Atmosphere	1 at	$= 1.013 \times 10^5 \text{ Pa}$
		$= 760 \text{ mm Hg}$
1 millimetre of mercury	1 mm Hg	$= 1.33 \times 10^2 \text{ Pa}$

Surface Tension

Perhaps you have seen the party trick of 'floating' a pin on the surface of water (Fig. 11.4) or you may have seen the pond insect which moves across the water's surface. Near to the point of contact the liquid surface is depressed. The surface supports the pin as though it were an elastic skin. This phenomenon of surface tension is a force acting in the surface which pulls it tight and tends to make the surface a minimum.

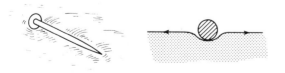

Fig. 11.4   Surface tension supports small objects

Capillary Action

The tension in the surface of a liquid in a tube causes it to be depressed or elevated. (Fig. 11.5.) The finer the tube the higher it is drawn. Plants and trees draw liquids to their upper parts by capillary action using their cells to form tubes of very small diameter. (Fig. 11.6.) Unfortunately the same action draws dampness up into building structures which do not have dampproof courses.

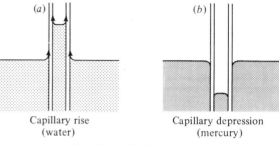

Fig. 11.5   Capillary action

$$\text{Rise} \propto \frac{1}{\text{tube diameter}}$$

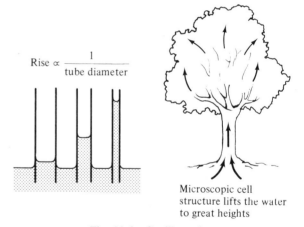

Microscopic cell structure lifts the water to great heights

Fig. 11.6   Capillary rise

Adhesion and Cohesion

Surface tension is a result of the force of attraction between the molecules of a liquid. These cohesive forces pull the surface molecules into the body of the liquid tending to make the surface as small as possible. (Fig. 11.7.) Thus liquid droplets tend to assume a spherical shape, i.e., the shape that has the least surface for a given volume. (Fig. 11.8.)

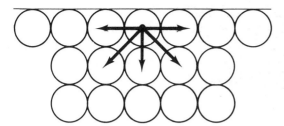

Fig. 11.7   Forces on a surface molecule result in a tension in the surface

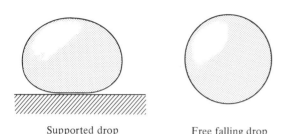

Supported drop   Free falling drop

Fig. 11.8   Liquid drops are pulled into a spherical shape by the surface tension

There is also a force of attraction between a liquid and any solid which contains it. This is called the force of adhesion.

Where cohesion is greater than adhesion a liquid clings more to itself than to the solid surface. It tries to form droplets and not to wet the solid, as mercury on glass. (Fig. 11.9.) Such a liquid has a convex surface in a capillary tube and is drawn down below the surrounding surface as in Fig. 11.5(b).

Where adhesion is greater than cohesion a liquid tends to cling to a solid surface. The liquid wets the solid (Fig. 11.9(a)) and in a capillary tube it forms a concave surface and is drawn up above the surrounding surface. (Fig. 11.5(a).)

Water-proofing lowers the adhesion of a fabric surface and the water rolls off as droplets. Detergents reduce the surface tension and have the opposite effect so that the water soaks in.

(a) Cohesion < adhesion   (b) Cohesion > adhesion
    water wetting a surface   mercury does not wet glass

Fig. 11.9   Cohesion and adhesion

Viscosity

Stirring a cup of tea needs no great effort but it requires some force. The liquid near the spoon is moved more quickly than the bulk of the liquid. The liquid layers are dragged over each other against the forces of cohesion we have described earlier. The property of a liquid which resists this relative motion is its viscosity. Motor oil is more viscous than tea and grease is very much more viscous than either.

The speed of a solid moving through a liquid is dependent on its viscosity, e.g., a boat on water and a gear wheel moving in oil. The rate of flow of liquid over stationary surfaces or through pipes also depends on the viscosity. The viscosity is measured in terms of a coefficient η. (Fig. 11.10.)

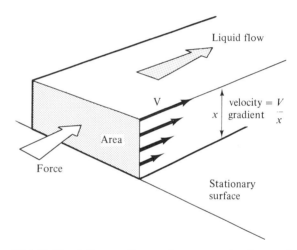

Fig. 11.10   A force acting on the cross-sectional area of a liquid establishes a transverse velocity gradient

Coefficient of viscosity

$$= \eta = \frac{\text{force}}{\text{area} \times \text{velocity gradient}}$$

Usually we are only interested in comparing viscosities and this can be done quite simply by comparing the times for a volume of liquid to flow through a standardized aperture. (See page 33.) Automotive oils are given a viscosity rating (SAE grade) by this method.

Viscosity is very much reduced as the temperature rises and oils may lose much of their lubricating properties at high temperatures. Great effort has been put into developing 'viscostatic' oils which flow freely at low temperatures but sustain their lubrication at high temperature.

Solutions

We tend to think of a solution as a solid dissolved in water like salt in the sea or sugar in tea, but this concept is too narrow. A solution is any mixture of substances in which the molecules of the substances are completely dispersed one among the other. Thus solutions can involve solids, liquids and gases.

Air is mainly a solution of oxygen and nitrogen. In fact all gases by their nature form solutions with each other. Whisky is basically alcohol dissolved in water while champagne is a solution of alcohol, water and carbon dioxide gas. (The characteristic flavours are due to other dissolved ingredients.) Rather more unfamiliar is the concept of a solid dissolved in another solid. Alloys are examples of this and so is glass and in each case the components are first brought together in the molten state.

In the case of a liquid solution the main liquid component is called the solvent and the minor component is called the solute, i.e., solvent + solute = solution; e.g., water + sugar = sugar solution. Certain solvents are able to dissolve a wide range of substances and are used in cleaning, e.g., water, carbon tetrachloride, acetone. Solvents which evaporate readily may be used as a means of carrying other substances as in paints or aerosols.

Although the molecules of the substances in solutions are interspersed they are not combined chemically and the proportions of the substances may vary. Dissolved substances cannot be separated by filters nor do they settle to the bottom when they are allowed to stand.

Solubility

In some solutions for example in gas mixtures or solutions of alcohol in water the proportion of one component may range from 0 to 100%. In other cases the proportion of one component reaches a maximum and the solution becomes 'saturated'. No more of that component will then dissolve in the solvent and any that is added will either fall to the bottom or remain suspended making the solution cloudy.

The solubility of a substance is the maximum proportion by mass which will dissolve in a particular solvent.

Thus for a saturated solution

$$\text{Solubility} = \frac{\text{mass of solute to saturate}}{\text{mass of solvent}} \text{ (kg/kg)}$$

The solubility is often quoted in grams per kilogram for convenience, i.e., as

$$\frac{\text{Mass of solute to saturate}}{\text{Mass of solvent}} \times 1000 \text{ (g/kg)}$$

For most solids the solubility increases with temperature. (Fig. 11.11.) There are exceptions, for

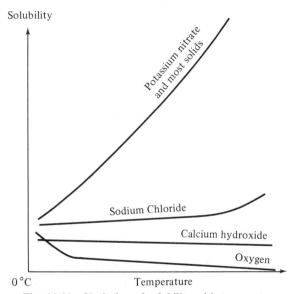

Fig. 11.11   Variation of solubility with temperature

example, the solubility of sodium chloride (common salt) remains constant while that of calcium hydroxide actually increases with temperature. The solubility of most gases reduces with increasing temperatures, for example, when water is heated the dissolved air bubbles out before it starts to boil. Pressure tends to increase the solubility of gases as you may observe when you uncork fizzy drinks. It may take some time for a solution to reach saturation. The rate at which substances dissolve is increased if the solute particles are very small and if the solution is heated and kept stirred. It also takes time for surplus solute to precipitate out from a super saturated solution.

An unsaturated solution is described by its concentration. This can be expressed as the number of grams of solute per kg of solvent. Needless to say the concentration reaches its maximum value and equals the solubility when the solution is saturated.

$$\text{Concentration} = \frac{\text{mass of solute}}{\text{mass of solvent}} \times 1000 \text{ (g/kg)}$$

Table 11.2 Solubilities of Common Substances at 25°C

Solvent	Solute	Solubility (g/kg)
Water	Sodium chloride (NaCl)	361
Water	Potassium hydroxide (KOH)	1190
Water	Copper sulphate ($CuSO_4$)	220
Water	Silver nitrate ($AgNO_3$)	2450
Water	Lead sulphate ($PbSO_4$)	0.045
Water	Oxygen (O_2)	0.04

Suspensions

Wood smoke contains very small solid particles suspended in the air. Each particle is big enough to contain millions of molecules but small enough to be more influenced by the movement of gas molecules than by the force of gravity. Although eventually the particles do settle the process may take a long time.

In a similar way powders form suspensions in liquids. When the particle size is large, e.g., sand, constant agitation is necessary to keep the suspension from settling to the bottom. As the particle size is reduced as in muddy water the suspension is maintained for a much longer period. When the particle size is very small, less than 10^{-6} m, i.e., about a thousand times the diameter of a molecule the settlement rate is so small as to be negligible. Such fine particles are called colloids. Suspended colloidal graphite is added to lubricating fluids to improve their effectiveness.

When colloids are suspended in liquids they pass right through filters and are very difficult to separate. For this reason they are referred to as 'colloidal solutions' although they are not true solutions.

Minute liquid droplets may also form suspensions. Mist is a suspension of water droplets in air. Where one liquid is suspended in another the result is called an emulsion, e.g., salad cream and plastic emulsion paint. Liquids like oil and water which do not normally mix can be made to form emulsions by ultrasonic vibrations. Old sump oil is a lovely mix, being an emulsion of oil and water produced by the heat and agitation of the engine. It also contains a suspension of dirt, gas and colloidal metal worn from the bearings which becomes corrosive, abrasive, and eventually lethal to engines. This abrasive effect is put to good use in special baths containing abrasives in suspension which when subject to high frequency vibration 'deburrs' the rough edges of metal components. (Fig. 11.12.)

Table 11.3 Types of Suspensions

Medium	Suspension	Result—example	
Liquid	Solid	Colloids	Aerosols
		Suspension	Paint
Liquid	Liquid	Emulsions	Salad cream
Liquid	Gas	Foams	Shaving lather
Gas	Solid	Smoke	Bonfires
Gas	Liquid	Mist	
Gas	Gas	(Form solutions not suspensions)	
Solid	Gas	Solid foams	Polystyrene

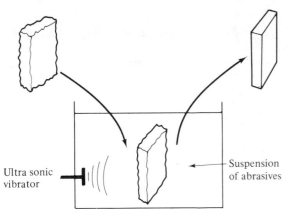

Fig. 11.12 Use of agitated suspension to 'deburr' metal stampings

Osmosis

When an aqueous solution is separated from pure water by a parchment barrier, something surprising happens. The water passes through the parchment thus diluting the solution. The surprising feature is that the process continues in the face of an adverse hydrostatic pressure (osmotic pressure) (Fig. 11.13.)

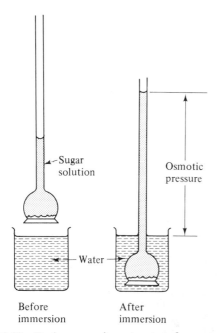

Fig. 11.13 During osmosis water transfers across the membrane to dilute the solution and establish a pressure difference

Osmosis also happens with other solutions with a suitable semi-permeable membrane. Water based solutions are most important because of their use by living organisms in absorbing and transporting essential substances. Plant roots absorb moisture from the soil when it is wet but retain the moisture when the soil is dry. The concentration of sugar in dried fruit draws water across the semi-permeable skin so that it swells to resemble its original condition. Prunes soaked in water try to become plums although they never quite make it.

The semi-permeable membrane allows the water to pass both ways but stops the larger sugar molecules. However, there is a greater concentration of water molecules on the water side than on the solution side (where they are mixed with sugar). The result is a net transfer of water to the solution.

Our kidneys purify our blood by using a semi-permeable membrane to concentrate the impurity which is then excreted. Cellophane is a synthetic semi-permeable membrane and has been used to form a temporary artificial kidney.

Questions

1. Why does the pressure of the atmosphere not crush us to death?

2. Why is it easier to cut through water than ice even though the water is more dense?

3. What evidence is there for thinking pitch is (a) a solid, (b) a liquid?

4. What physical processes transport water from the roots to the branches of a tree?

5. Why is the surface of a liquid horizontal?

Multiple Choice

1. Which statement is **not** true of liquid pressure?
   (a) increases with depth,
   (b) the same at all depths,
   (c) acts at right angles to container wall,
   (d) varies with density.

2. The pressure at a depth in a liquid depends on which factor?
   (a) surface tension, (b) viscosity, (c) density, (d) depth.

3. Which is **not** a unit of pressure?
   (a) pascal, (b) mm Hg, (c) newton per square metre,
   (d) kilogram per cubic metre.

4. Which process causes dampness in walls without dampproof courses?
(a) osmosis, (b) adhesion, (c) cohesion, (d) capillary action.

5. Which is the most viscous?
(a) grease, (b) petrol, (c) oil, (d) water.

6. In which are the components more finely dispersed?
(a) suspension, (b) colloid, (c) solution, (d) mixture.

7. What is the name given to a liquid which dissolves another substance?
(a) solution, (b) solute, (c) solvent, (d) suspension.

8. What is the name given to a solution that will not dissolve any more solute?
(a) saturated, (b) dissolved, (c) suspended, (d) super saturated.

9. Which is a solid solution?
(a) mixture, (b) suspension, (c) compound, (d) alloy.

10. Which is **not** a suspension?
(a) smoke, (b) air, (c) mist, (d) emulsion paint.

Problems

1. What pressure is exerted when a force of 2000 N is applied over a surface area of 0.1 square metre? [20 kPa]

2. A force of 40 N is applied perpendicularly to a rectangular area measuring 12 mm × 20 mm. What is the pressure in pascals? [167 kPa]

3. Calculate the pressure on a brick of mass 1.8 kg and measuring 300 mm × 100 mm × 50 mm when it is stood on each of its surfaces. (The acceleration due to gravity = 10 m/s^2.) [600 Pa; 1200 Pa; 3600 Pa]

4. If a water pressure of 262 kilo pascal (kN/m^2) is required in a water main pipe, how much higher must the storage tank be? (Density of water = 1000 kg/m^3; acceleration due to gravity is 9.8 m/s^2.) [26.7 m]

5. What pressure is exerted at the base of a column of mercury 760 mm high? The density of mercury is 13 600 kg/m^3 and the acceleration due to gravity is 9.8 m/s^2. [101 kPa]

6. Calculate the water pressure on a submarine at a depth of 200 m below sea level. The density of sea-water is 1025 kg/m^3 and $g = 9.8$ m/s^2. [2.01 MPa]

7. What mass of water is contained in a tank of base dimensions 1.0 m × 2.0 m filled to a height of 1.5 m? Density of water 1000 kg/m^3. What force does it exert on the base? ($g = 9.8$ m/s^2.) What pressure is exerted on the base? [3000 kg; 29 kN; 15 kPa]

8. What height of column of water would exert the same pressure as a column of mercury 760 mm high? (Density of water = 1000 kg/m^3; density mercury 13 600 kg/m^3.) [10.3 m]

9. Calculate the solubility of copper sulphate if 110 g of it is just sufficient to saturate 0.50 kg of water. 220 g/kg? [837 g]

10. 250 g of a saturated sodium chloride solution is heated until all the water is evaporated off. How much salt will remain? The solubility of sodium chloride salt is 361 g/kg. [66.3 g]

11. The concentration of a solution of copper sulphate is 25 g/kg. How much salt does 4.4 kg of the solution contain? [107 g]

12. How much salt would be required to bring the solution in problem 11 up to a saturated solution of 220 g/kg? [837 g]

13. 400 g of water is added to 500 g of a saturated solution of copper sulphate. Calculate the resulting concentration of this solution if the solubility of copper sulphate is 220 g/kg. (Watch out it's tricky.) [111 g/kg]

Multiple Choice Answers

1(b), 2(cd), 3(d), 4(d), 5(a), 6(c), 7(c), 8(a), 9(d), 10(b).

12.
Gases and Vapours

Nearly all the matter in the universe is gaseous. The stars, which constitute most of this matter, are made of gas though in their interiors the density exceeds that of any solid we know on earth. Across some areas of space stretch great clouds of gaseous matter, whose density is less than the most complete vacuum ever produced on earth, and whose temperature is close to absolute zero.

Although the chemical properties of gases vary, their physical properties are very similar. They all spread themselves rapidly to every corner of a space into which they are introduced. They are all transparent and usually colourless. Gases mix with each other very readily and quickly get to a constant composition throughout the space they occupy.

A few simple laws, recalling the names of the scientists who formulated them, sum up the characteristic behaviour of gases.

The Gas Laws

The gas laws express relations between the volume V, pressure P, and temperature T of a fixed mass of gas.

General Gas Law **The product of the pressure and volume of a fixed mass of gas is proportional to its absolute temperature, i.e.,**

$$PV \propto T$$

The law is sometimes stated as three separate relations for a fixed mass of gas.

Boyle's Law

$$P \propto \frac{1}{V} \text{ at constant temperature}$$

For example, if the pressure on a gas is doubled its volume is halved. (Fig. 12.1.) At extremely low pressures all gases obey the law, closely. (Fig. 12.2.) At higher pressures deviations may become significant.

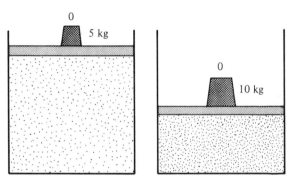

Fig. 12.1 A specific example illustrates the meaning of Boyle's law

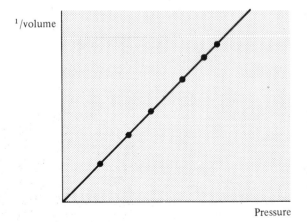

Fig. 12.2 Graph of 1/volume against the pressure of a fixed mass of a gas at constant temperature (Boyle's law)

Charles' Law

$V \propto T$ at constant pressure (Fig. 12.3)

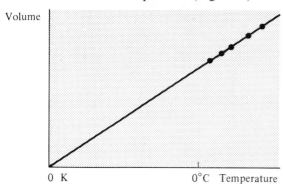

Fig. 12.3 Graph of volume against absolute temperature of a fixed mass of gas (Charles' law $V \propto T$)

Pressure law

$P \propto T$ at constant volume (Fig. 12.4)

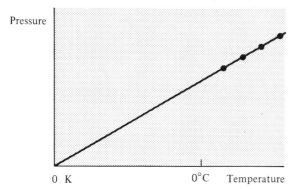

Fig. 12.4 Graph of the pressure against the absolute temperature of a fixed mass of gas (Pressure law $P \propto T$)

We may express the combined effect of the laws as an equation

$$PV = k\ T$$

where k the constant of proportionality in this equation, applies to a fixed mass of a particular gas.

Expressing this as

$$\frac{PV}{T} = k$$

we see that

$$\frac{PV}{T}$$

is a constant for a fixed mass and when a gas changes from one set of conditions P, V, T, to another set of conditions $P_2 V_2 T_2$ we can write

$$\frac{P_1 V_1}{T_1} = \frac{P_2 V_2}{T_2}$$

A common use of the equation is to convert a measured volume of gas to the volume it would occupy under standard conditions of temperature and pressure (S.T.P.). The standard conditions of temperature and pressure are 0°C and 760 mm of mercury pressure. This pressure, exerted by a column of mercury 760 mm high, is referred to as one atmosphere.

$$1 \text{ atmosphere} = 1.013 \times 10^5 \text{ pascals (N/m}^2)$$

Problem

The density of oxygen at S.T.P. is 1.43 kg/m³. What mass is contained in a cylinder 1.4 m long and internal diameter 0.3 m if it exerts a pressure of 12 atmospheres at 15°C.

Solution

We change the volume to the equivalent volume at S.T.P. using

$$\frac{P_1 V_1}{T_1} = \frac{P_2 V_2}{T_2}$$

$$\frac{12 \times \pi \times (0.15)^2 \times 1.4}{288} = \frac{1 \times V_2}{273}$$

$$V_2 = \frac{273}{288} \times \frac{12}{1} \times \frac{\pi}{4} 0.3^2 \times 1.4 = 1.125 \text{ m}^3 \text{ at S.T.P.}$$

$$\begin{aligned}
\text{Mass} &= V_2 \times \text{density at S.T.P.} \\
&= 1.125 \times 1.43 \\
&= 1.61 \text{ kg}
\end{aligned}$$

We can also use the gas equation to determine the mass of a gas in a container by measuring the pressure, temperature, and volume of the gas. Although it is possible to measure the mass directly by weighing the container, first under vacuum and then filled with gas, this method is usually extremely inconvenient. An oxygen cylinder, for example, is a difficult object to weigh in this manner, whereas measuring the pressure and temperature of the gas inside it is easy.

Dalton's Law In a mixture of gases, the total pressure is the sum of the pressures which each gas would exert if it alone occupied the total available volume.

$$p = p_1 + p_2 + p_3, \text{ etc.}$$

The law just confirms what we would expect. It means that we can predict the overall effect in a mixture of gases by simply adding the separate contributions.

Avogadro's Law

The law states that **at a given temperature and volume, the pressure exerted by any gas depends only on the number of molecules present.** Neither the shape nor the size of the molecules of a gas influences the pressure they exert. Thus, the pressure exerted by a million molecules of hydrogen in a given container would be the same as that exerted by a million molecules of oxygen, or of chlorine, or of radon. Avogadro's law is sometimes stated as: 'Equal volumes of all gases under the same conditions of temperature and pressure contain the same number of molecules'. With a little thought you will see that the two statements of the law are equivalent. In calculations we often use the fact that the molecular mass in grams of any gas (6.02×10^{23} mols) occupies 22.4 litres at S.T.P.

Problem
What volume would 20 g of helium occupy at S.T.P. (mol mass helium = 4)?

Solution
4 g of helium occupy 22.4 litres.

$$\therefore \ 20 \text{ g of helium occupy } \frac{20}{4} \times 22.4 = 112.0 \text{ litres}$$

Vapours

In the minds of most people, the word 'vapour' conjures up an image far removed from the formal scientific meaning of the word. The mist that hangs over hollows in the early morning, for instance, is not a vapour, nor is the smoke that curls above volcanic craters, nor is the steam that rises from boiling liquids. These are in reality suspensions of liquid droplets.

Vapours are invisible, as gases are, and they resemble gases in many other ways. Between the two states of matter, however, there are essential differences in behaviour which oblige us to recognize an essential difference in kind. For example, when a vapour is compressed, its pressure rises until it reaches a point at which the vapour condenses into a liquid. On the other hand, when a gas is compressed, it remains a gas, even when subject to very high pressures.

Evaporation

According to the kinetic theory of matter, the molecules within a liquid are continually on the move. Some molecules, which by chance collisions gain a velocity, much higher than average, shoot across the surface of the liquid and escape to become vapour. The higher the temperature of the liquid, the greater the velocity of the molecules, and the greater the rate of evaporation.

If the space above the liquid is not bounded by a container, the molecules which escape keep right on going and diffuse away. Evaporation continues until all the liquid is gone. The process is accelerated if there is a movement of air above the liquid.

Exposed liquids are usually a few degrees below the temperature of their surroundings because of the extraction of latent heat of evaporation. After an initial drop in the temperature of the liquid, heat flows in from the environment at the same rate as it is lost by evaporation.

Saturated Vapour

If the space above a liquid is enclosed then vapour molecules accumulate in the space. Occasionally, molecules from the vapour meet the liquid surface and return to the liquid. As the density of the vapour increases, a stage is reached when the number of molecules evaporating is equal to the number of molecules condensing in the same time. (Fig. 12.5.) Under these conditions, the space above the liquid is said to be saturated with the vapour from the liquid.

The vapour molecules exert a pressure which increases as the density of the vapour increases. When the vapour is saturated, the pressure exerted

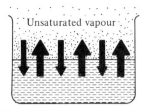 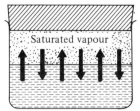

Fig. 12.5 In a saturated vapour molecules leave and rejoin the liquid at the same rate

reaches a steady value, called the saturated vapour pressure or S.V.P.

A vapour can exist when it is too cold for the liquid to exist, i.e., below the freezing point. In the polar regions of the earth, the water vapour evaporates from snow and ice and falls again as snow. This immediate change from solid to vapour, which takes place without an intermediate liquid state, is called sublimation.

Relation Between Pressure and Boiling Point

Imagine a bubble of vapour near the surface of a liquid. (Fig. 12.6.) The pressure tending to collapse the bubble is the same as that exerted on the surface, i.e., atmospheric pressure. The pressure tending to expand the bubble is the S.V.P. of the liquid. When the S.V.P. exceeds atmospheric pressure the bubble expands and releases the vapour above the surface, i.e., the liquid boils. Thus, a liquid at a uniform temperature boils first at the surface, where the pressure is least.

Fig. 12.6 The pressure of the saturated vapour in a bubble tends to expand it while the pressure of the liquid tends to contract it

Water, as we know, boils at 100°C when atmospheric pressure is at the standard value of 76 cm of mercury. At high altitudes, where the atmospheric pressure is below the standard value, the temperature at which the S.V.P. of water equals atmospheric pressure is less than 100°C. For example, at the top of Mount Everest, where atmospheric pressure is only 230 mm Hg, water boils at 70°C, so that, in addition to being unable to breathe enough air up there, you cannot make a decent cup of tea or coffee. For the same reason, a puncture in a spaceman's suit would be disastrous, because the pressure in space is almost zero and the temperature of his body is well above the boiling point of the body fluids.

Vapours and Gases

Are substances divided fundamentally into two groups, those which form vapours and those which form gases? This question is best answered by looking at the results of an experiment carried out by Andrews. He subjected carbon dioxide to a wide range of pressure, volume and temperature changes and presented his results on a pressure–volume graph. (Fig. 12.7.)

Starting with the carbon dioxide at one of the temperatures on the right of Fig. 12.7 he reduced the volume and measured the resulting pressure while the temperature was kept constant. The lines are called isothermal curves.

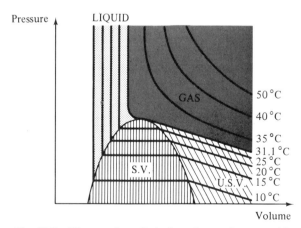

Fig. 12.7 The results of Andrews' experiments with carbon dioxide plotted on a $P - V$ diagram

129

Critical Temperature

One particular isothermal curve marks the difference between liquid and gas. This isothermal is the first one to have a horizontal section. The isothermal defines the critical temperature of the substance. Above the critical temperature a gas cannot be liquefied by increased pressure alone. No increase in pressure will cause the gas to change into a liquid. It merely becomes a denser gas. Below the critical temperature a substance may exist as a solid, a liquid or a vapour. The isothermals below the critical temperature, at the bottom right of Fig. 12.7 represent the unsaturated vapour. No liquid is present at this stage and the isothermals are like those for the gas. An unsaturated vapour obeys the gas laws and is indistinguishable from a gas until it is subjected to high pressures, in which case the vapour condenses.

The isothermals for the saturated vapour are horizontal. During this stage, as the volume reduces, the pressure remains constant and the liquid condenses. When all the vapour has been condensed to liquid, the isothermals become nearly vertical; this indicates that the liquid state is almost incompressible. Any substance will behave like carbon dioxide as represented in Fig. 12.7 over an appropriate range of pressures and temperatures, and provided, of course, the substance does not decompose. The critical temperatures of some substances are given in Table 12.1.

Table 12.1   Critical Temperatures

Element	K	°C
Hydrogen	33	− 240
Oxygen	154	− 119
Nitrogen	126	− 147
Air	133	− 140
Water	647	374
Aniline	699	426

The Kinetic Theory of Gases

Bernoulli a Swiss scientist first suggested the theory that gases were composed of particles in rapid motion.

The theory was supported by observation of minute smoke particles under a microscope. The particles were observed to make small jerky movements called Brownian motion. Because the smoke particles are small the random collisions of the gas molecules produced a visible movement. Larger particles do not exhibit the effect because they experience more collisions per second which tend to balance out and any imbalance that does arise has an imperceptible effect on the greater mass. (Fig. 12.8.) The gas laws can be explained in detail by assuming the kinetic theory and so they provide evidence to support the theory.

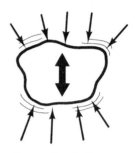

Fig. 12.8   Brownian motion is due to a momentary imbalance of forces exerted by random collisions. In one second, one-sixth of all the molecules within a distance, c, of the surface strike the surface.

The continual bombardment of any surface by the gas causes a pressure to be exerted. The greater the density of a gas, the more frequent the bombardment and the greater the pressure exerted. Thus, the pressure increases either when more gas is pumped into a vessel, or when the volume of a certain mass of gas is reduced. When the temperature of a gas is raised, the speed of the molecules increases causing an increase in both the number and the momentum imparted by each collision. This accounts for the increase in pressure of a gas with temperature. The kinetic theory is a concept which applies not only to gases but also to liquids and solids. The agitation of atoms in a solid is rather like a tightly packed crowd standing on the terraces at a football match, having some pattern and order but being difficult to penetrate. The liquid state is comparable with a crowded disco; more energy, more movement but smooth and orderly. The gaseous state resembles a school playground. Many children make violent and frequent collisions with each other and the boundary. This mental picture may be limited but it is vivid and useful when considering the different properties of solids, liquids and gases.

Pumps

A pump is a mechanical device designed for transferring a fluid from one place to another by creating a difference in the pressures within that fluid. The most familiar type of pump to most of us is the kind used to inflate tyres. (Fig. 12.9.)

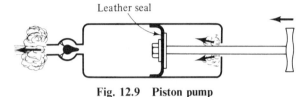

Fig. 12.9   Piston pump

This consists of a cylinder, inside which a plunger moves backwards and forwards, compressing air with each forward stroke, and forcing it into the tyre through a valve, which closes automatically with every return stroke of the plunger. When the pliable leather washer on the plunger is reversed, the pump can be used in conjunction with an appropriate valve to produce a vacuum.

The water filter pump is a device for pumping a gas which uses something available in most places, a supply of water under pressure. Water passes rapidly across the end of a tube and knocks away any molecules of the gas which emerge. This continually extracts the gas molecules, even when the pressure of the gas inside the tube is a fraction of atmospheric pressure. (Fig. 12.10.)

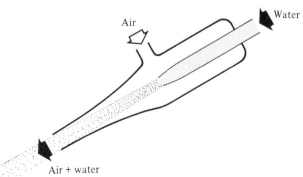

Fig. 12.10   Water diffusion pump

Pressure Measurement

Pressures up to several times greater than atmospheric pressure can be measured by the Bourdon gauge. Basically, the gauge is a curved bronze tube whose curvature varies with the pressure. (Fig. 12.11.) The movement is magnified mechanically

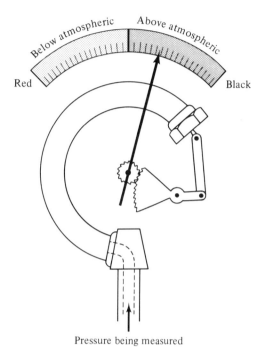

Fig. 12.11   The Bourdon gauge

and registers on a scale calibrated directly in pressure. Pressures in excess of atmospheric pressure—counting atmospheric as zero—are referred to as gauge pressures to distinguish them from absolute pressures.

$$\text{Absolute pressure} = \text{Gauge pressure} + \text{Atmospheric pressure}$$

Thus the gauge pressure is zero when a system is at the same pressure as the atmosphere.

Moderate pressures comparable with atmospheric pressure can be measured by a U-tube containing a liquid. (Fig. 12.12.) When one arm is open to the atmosphere it registers gauge pressures. If one arm is closed and the space above the liquid evacuated of air the device records absolute pressures. Atmospheric pressure itself is measured by the fortin barometer which is really a

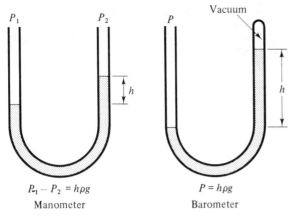

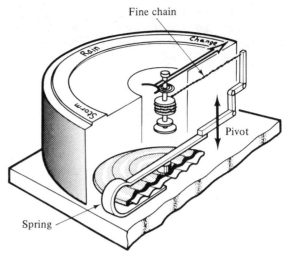

$$P_1 - P_2 = h\rho g$$
Manometer

$$P = h\rho g$$
Barometer

Fig. 12.12 The U-tube manometer measures pressure differences; where one liquid column is bounded by a vacuum it measures absolute pressure

Fig. 12.14 Aneroid barometer movement. A sealed disc-shaped container is partially evacuated. As it expands and contracts with changes in atmospheric pressure this movement is magnified mechanically to give a visible reading

modified U-tube. (Fig. 12.13.) The aneroid barometer, though less accurate, is more convenient. (Fig. 12.14.) Day to day variations of the barometer readings help us to predict weather while the regular reduction of atmospheric pressure with altitude allows the aneroid barometer to act as an altimeter.

A continuously reading gauge used extensively to measure very low pressures from 10^{-1} to 10^{-4} mm Hg is the Pirani gauge. It is based on the regular variations of the thermal conductivity of a gas with pressure. The conductivity of a gas affects the equilibrium temperature of a heated filament which, in turn, affects its electrical resistance. This is converted electrically to a deflection on a meter. (Fig. 12.15.)

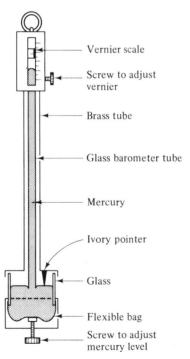

Fig. 12.13 Fortin barometer. Before reading the atmospheric pressure on the vernier, the mercury level is set to touch the ivory pointer

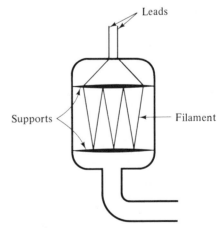

Fig. 12.15 The Pirani gauge measures pressure by its effect on thermal conductivity at low pressures

Pressures as low as 10^{-13} mm Hg have been produced for the purpose of examining physically clean surfaces, i.e., surfaces completely free from absorbed substances. At the other extreme, pressures up to 400 000 atmospheres can be achieved. Astonishing effects occur at such high pressures, such as a 50% reduction in a volume of water and an increase in the viscosity of machine oil to approach that of pitch.

Humidity

Animals and plants are composed mainly of water and depend on an interchange of moisture with the atmosphere. Sweating in mammals regulates body temperature. Transpiration in plants is a process whereby water is drawn in through the roots and evaporated at the leaves. Some industrial processes, such as cotton-spinning and paint-drying, are strongly affected by atmospheric moisture. The properties of certain organic materials, e.g., paper, vary with humidity so that control is necessary when testing or manipulating them.

The mass of water vapour contained per unit volume of air is the absolute humidity. Surprisingly, these processes are not influenced by the absolute humidity but by how close the air is to saturation. This is measured by the relative humidity, defined for a given volume, as follows:

$$\text{R.H.} = \frac{\text{mass of water vapour present}}{\text{mass required to saturate at same temperature}} \times 100$$

The relative humidity is usually referred to simply as 'humidity' and is expressed as a percentage. A 100% humidity corresponds to saturation. The mass of water vapour required to saturate the air varies with temperature. (Table 12.2.)

The humidity can be increased either by the evaporation of more water or by a reduction in temperature. Thus, air approaches nearer to saturation as it cools. Eventually the air becomes completely saturated and water condenses into droplets which either remain suspended in the air or form on exposed surfaces as dew. This condensation begins at the dewpoint, i.e., the temperature at which the water vapour present is just sufficient to saturate the air. Any cooling below the dewpoint produces further condensation.

Table 12.2   Mass of Water Vapour Contained in One Cubic Metre of Saturated Air

Temperature (°C)	Water vapour content (g/m³)
0	4.84
5	6.76
10	9.33
15	12.71
20	17.12
25	22.84
30	30.04
35	39.18

The Wet and Dry Bulb Hygrometer

The wet and dry bulb hygrometer is the most convenient instrument for measuring humidity. The method is an empirical one, that is, it depends on no exact theory. In the instrument, two identical mercury thermometers are mounted side by side. The bulb of one thermometer is enveloped by wetted fabric which has the same effect on the thermometer as wet clothing has on a person. It makes it cooler. The dryer the air, the more rapid the evaporation and the greater the depression of the wet thermometer reading below that of the dry thermometer. For a given dry bulb temperature, the difference in temperature of the two thermometers correlates with the humidity of the air. Sometimes, the thermometers are mounted in a sling which is rotated rather like a football rattle, before readings are made. (Fig. 12.16.)

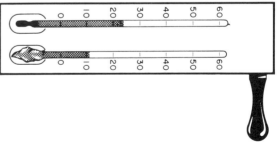

Fig. 12.16   The sling hygrometer (psychrometer) used to measure relative humidity ensures a rapid replacement of air near the thermometer bulbs

Alternatively they are mounted on a bracket and the water is drawn onto the wet bulb by capillary action from a container. Each particular wet and dry bulb hygrometer has its own empirical tables from which the actual humidity is obtained.

For the recording and control of humidity in industry, hygrometers depending on the length of an animal hair or the amount of curl in a paper spiral are used. (Fig. 12.17.) They are convenient because they give a direct reading of humidity, but they need frequent recalibration.

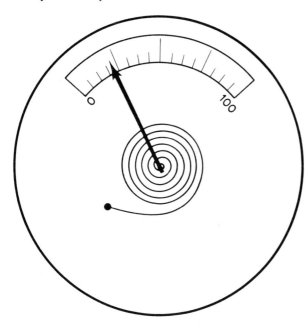

Fig. 12.17   The paper strip hygrometer uncurls when the humidity is high, giving a direct reading of humidity

The humidity of the air has an effect on how warm the air feels to the human body. Humid air always feels warmer than dry air of the same temperature. Some respiratory conditions, such as bronchitis and hay fever, may be relieved by high humidity. The central heating of our homes is rapidly replacing older and less effective methods of heating but the homes of the future will be air conditioned which, in addition to heating involves cooling, ventilating and humidifying.

Questions

1. Why do solids, liquids and gases behave differently when subject to a vacuum?

2. Could steam have a greater density than ice?

3. How could you show the difference between a gas and a vapour?

4. Why is humidity so important for processes involving organic matter?

5. What stops the atmosphere from escaping into the vacuum of space?

Multiple Choice

1. The volume of a fixed mass of gas is proportional to which two factors?
(a) pressure, (b) absolute temperature,
(c) $\dfrac{1}{\text{pressure}}$, (d) $\dfrac{1}{\text{absolute temperature}}$.

2. Standard temperature is
(a) 0 K, (b) 100°C, (c) 373 K, (d) 0°C.

3. Standard pressure is
(a) zero pressure, (b) 760 mm Hg,
(c) 100 mm Hg, (d) 1 pascal.

4. What is the effect of increasing the pressure of a gas?
(a) evaporation, (b) drop in temperature,
(c) volume increase, (d) volume decrease.

5. The rate at which water evaporates depends on
(a) draught, (b) mass, (c) surface area, (d) humidity.

6. Above the critical temperature a substance may exist only as
(a) a vapour, (b) a liquid, (c) a gas, (d) a solid.

7. The temperature at which a liquid boils can be raised by increasing
(a) heat supply, (b) pressure, (c) impurity,
(d) size of container.

8. Temperature increase of a gas at constant volume is accompanied by an increase in molecular
(a) size, (b) separation, (c) speeds, (d) energy.

9. The effect of evaporation on the liquid remaining is to decrease its
(a) mass, (b) density, (c) temperature, (d) heat.

10. A hygrometer is a device for measuring
(a) temperature, (b) humidity, (c) pressure,
(d) density.

Problems

1. Estimate the force exerted by atmospheric pressure (10^5 Pa) on a television tube screen measuring 400 mm × 500 mm. [2×10^4 N]

2. A closed vessel contains a gas at 1 atmosphere pressure. If the volume of the vessel is doubled without any change in temperature, calculate the new gas pressure. [$\frac{1}{2}$ atmosphere]

3. The volume of a mass of hydrogen is 3.5 m³ at 15°C and 1000 Pa pressure. Calculate the volume at 0°C and 100 000 Pa pressure. [0.033 m³]

4. A metal cylinder containing a gas at 12 atmospheres pressure is taken from a cold store at -23°C. Calculate the new pressure in atmospheres when the cylinder is removed from the store and acquires the surrounding temperature of 27°C. [14.4 atmospheres]

5. What mass of oxygen is contained in a cylinder 1.2 m long and of internal diameter 0.20 m if it exerts a pressure of 5.6 atmospheres at 20°C? (Density of oxygen at 1.0 atmosphere and 0°C = 1.43 kg/m³.) [0.28 kg]

6. An electric light bulb is filled with argon at 100 mm Hg pressure at 15°C. What is the gas pressure when the lamp is alight, at 400°C? [234 mm Hg]

7. At 15°C and 750 mm Hg pressure, air weighs 1.26 kg/m³. What is its density at 250 mm Hg pressure and -45°C? [0.53 kg/m³]

8. A car tyre is inflated to 28.5×10^4 Pa at 12°C. Running on the tyre raises its temperature to 27°C. What is its pressure? [3.0×10^4 Pa]

9. A kilogram of air at 0°C and 1 atm pressure is heated to 100°C at constant pressure. What is the volume of the heated air? What pressure would be needed to bring the air back to its original volume if it remains at 100°C? (Volume of 1 kg of air at 0°C and 1 atm pressure = 0.774 m³.) [1.06 m³, 1.37 atm]

10. A gas has a density of 1.35 kg/m³ at S.T.P. (273 K and 1.01×10^5 Pa. What would be the mass of 1.0 cubic metre of the gas at a temperature of 280 K and a pressure of 3.0×10^5 Pa? [3.91 kg]

11. Gas at 20 atmospheres pressure is stored in a cylinder of internal dimensions 0.30 m long × 0.12 m diameter. Calculate its volume in litres. It is released into a plastic bag at 1 atmosphere pressure. What volume of the bag is inflated if the temperature stays constant? [3.4 litres, 68 litres]

Multiple Choice Answers

1(bc), 2(d), 3(b), 4(d), 5(acd), 6(c), 7(bc), 8(cd), 9(acd), 10(b).

13.
Chemical Reactions

Perhaps like many you find chemistry is daunting to the memory. This chapter deals with a few basic processes in terms of those chemicals that technicians most often meet. Learning just a few names of chemicals and their formulae would repay your effort as a basis to an understanding of engineering materials and processes.

All matter is made up of only about 100 different elements. The very much larger number of different materials that we meet are explained by the different chemical combination of the elements that can exist. When elements combine chemically, the product may bear no relation to its component parts. For example two gases oxygen (O) and hydrogen (H) combine chemically to produce a liquid, water. 2 molecules of hydrogen + 1 molecule of oxygen = 2 molecules of water.

$$2H_2 + O_2 = 2H_2O$$

Every molecule of a particular compound contains exactly the same number of atoms of the same elements. When a compound is being made from its elements they always take part in the same ratio by weight. This is the law of constant proportion. If there is a surplus of any component it is left over at the end and does not take part in the reaction. Thus pure water always has the same chemical composition [H_2O] and the proportion of oxygen and hydrogen does not vary.

Mixtures and Compounds

It is important to distinguish between a physical mixture and a chemical compound. In a mixture, the components make only superficial contact, retaining their separate identity, and being capable of separation from the mixture by physical means. A mixture has something of the properties of its components and its percentage composition may vary. A mixture of peas and beans illustrates these points well.

A compound, formed of two or more elements, involves a much more fundamental change. The combination is accompanied by the absorption or emission of heat, and the properties of the compound produced may be quite different to those of either component. For example, hydrogen and oxygen are quite unlike water, which is the compound they produce when they combine chemically. The proportions of the components are constant in every part of a compound, and separation can only be achieved by chemical means. (Table 13.1.)

The difference between a compound and a mixture can be illustrated very neatly with iron filings and powdered sulphur. These two components mixed in any proportion can be separated by using

Table 13.1 The Differences Between Mixtures and Compounds

Mixtures	Compounds
1. The substances can be separated physically	The substances can only be separated by chemical change
2. The composition can vary	The composition is constant
3. The properties of a mixture are the sum of the properties of its parts	The properties of a compound differ from those of its components
4. There is usually no production of heat on mixing	Formation involves the evolution or absorption of heat

a magnet or by placing the mixture in a liquid in which the sulphur either floats or dissolves. If the mixture is heated, however, a compound of iron and sulphur, iron sulphide, is formed which has properties quite unlike those either of the iron or of the sulphur. The compound has fixed proportions of iron and sulphur, any surplus of either in the mixture being left unchanged. (Fig. 13.1.)

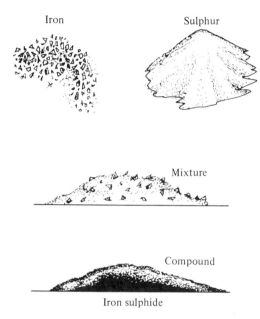

Fig. 13.1 **The difference between a mixture and a compound**

The physical separation of mixtures although possible can often be difficult (tar and feathers?). Some standard methods are shown in Fig. 13.2.

Chemical Symbols

Every element has a symbol which represents it and provides a shorthand way of writing it down. In many cases the symbol is just the capital initial letter of the element, e.g., C represents a carbon atom and S represents a sulphur atom. Where elements have the same initial then a smaller second letter is used to distinguish them, e.g.,

C represents carbon
Cl represents chlorine
Co represents cobalt

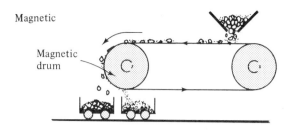

Magnetic

Magnetic drum

Filtration

Non dissolving

Evaporation

Solution   Solid

Distillation separates high boiling point liquid

Separation of immiscible liquids

Lighter liquid

Heavier liquid

Crystallization

Evaporation from a saturated solution produces crystals of the solute

Fig. 13.2 **Separation of mixtures by physical means**

Sometimes the letters from the latin name of the element are used, e.g.,

Sodium is represented by Na from its Latin name Natrium

Table 13.2 gives some elements and their symbols.

Table 13.2   Chemical Symbols

Metals	Symbol	Non metals	Symbol
Potassium	K	Hydrogen	H
Sodium	Na	Oxygen	O
Aluminium	Al	Sulphur	S
Magnesium	Mg	Carbon	C
Zinc	Zn	Chlorine	Cl
Iron	Fe	Nitrogen	N
Copper	Cu		
Silver	Ag		

An atom is the smallest part of an element that can take part in a chemical change. But atoms are not always stable when they are on their own. Some atoms form pairs or trios which move together as units. For example hydrogen gas consists of pairs of atoms moving freely about. The stable unit is called a molecule and the number of atoms it contains is represented by a suffix.

H represents an atom of hydrogen
H_2 represents a molecule of hydrogen

This combining of atoms of the same element into stable groups only happens for some gases. Most of the elements have atoms which are quite stable when they are single, i.e., the molecule contains only one atom. Apart from these gases we use the word atom to describe the smallest part of an element and the word molecule to describe the smallest part of a compound. (Table 13.3.)

Table 13.3

Element	Atom	Molecule	Description
Sulphur	S	S	monatomic
Carbon	C	C	monatomic
Hydrogen	H	H_2	diatomic
Nitrogen	N	N_2	diatomic
Chlorine	Cl	Cl_2	diatomic
Oxygen	O	O_2	diatomic
		O_3	triatomic

The triatomic form of oxygen is rare and has the special name of Ozone

The chemical symbols describe compounds better than words because they indicate clearly the number of atoms of each constituent. (Table 13.4.)

Table 13.4   Symbols for Some Compounds

Ferric sulphate	FeS
Carbon monoxide	CO
Carbon dioxide	CO_2
Water	H_2O
Hydrogen sulphate (sulphuric acid)	H_2SO_4
Hydrogen chloride (hydrochloric acid)	HCl
Sodium chloride (common salt)	NaCl
Sodium hydroxide (caustic soda)	NaOH

Chemical Equations

Using the chemical symbols we can represent chemical reactions by equations in a very clear and concise way:

$$Iron + Sulphur = Iron\ sulphide$$
$$Fe + S = FeS$$
$$Carbon + Oxygen = Carbon\ dioxide$$
$$C + O_2 = CO_2$$

Copper oxide is decomposed when heated in hydrogen

$$CuO + H_2 = Cu + H_2O$$

i.e., metallic copper and water are formed

$$Magnesium + Oxygen = Magnesium\ oxide$$
$$2Mg + O_2 = 2MgO$$

In writing chemical equations like this remember that the large prefix numbers indicate the number of molecules and the small suffixes indicate the number of atoms normally in the molecule. Thus $2H_2$ indicates two molecules of hydrogen each of which contains 2 atoms, and $2H_2O$ indicates 2 molecules of water each of which contains 2 atoms of hydrogen and one of oxygen. Note also that the equations balance, i.e., the number of atoms of an element on one side equal the number on the other. We illustrate this by writing the equation for the reaction when hydrogen burns to form water.

$$H + O = H_2O$$

To make it balance we must have 2 hydrogen atoms on the left hand side, i.e.,

$$2H + O = H_2O$$

This equation describes the reaction well enough but hydrogen and oxygen are both diatomic and it is better to deal in molecules

$$2H_2 + O_2 = 2H_2O$$

Problem

Write a balanced equation for the reaction between sodium (Na) and the diatomic gas chlorine (Cl).

Solution

$$Na + Cl = NaCl$$

But Cl should be written Cl_2 to describe a molecule

$$Na + Cl_2 = NaCl$$

But we must have 2 chlorine atoms on the right hand side to match the Cl_2

$$Na + Cl_2 = 2NaCl$$

Now we must have 2 sodium atoms on the left to balance those on the right, hence

$$2Na + Cl_2 = 2NaCl$$

Exothermic Reactions

When a chemical reaction takes place heat is usually given out. For example the basic reaction in a coke fire is the oxidation of carbon, i.e., the combination of carbon and oxygen to give the gas carbon dioxide.

$$C + O_2 = CO_2 + heat$$

We may need a little heat to get it going but after that it is self sustaining. Heat releasing reactions like the burning of carbon are described as exothermic. The products of highly exothermic reactions tend to be stable. Carbon dioxide is an inert gas.

Combinations of substances which absorb heat are called endothermic reactions and they are less common. For example carbon combines endothermically with sulphur and absorbs heat

$$C + 2S + heat = CS_2 \text{ (carbon disulphide)}$$

The products of endothermic reactions tend to be unstable. Carbon disulphide is highly inflammable

and for good measure it is poisonous and probably has a more nauseating smell than any other substance.

Chemical Radicals

Certain chemical combinations act as a self contained unit in combining with other elements. For example one atom of sulphur and four atoms of oxygen form the sulphate unit SO_4. This unit or radical combines with other elements, e.g.,

$CuSO_4$ Copper sulphate
H_2SO_4 Hydrogen sulphate (sulphuric acid)
$ZnSO_4$ Zinc sulphate

Another example is the nitrate radical NO_3 which is present in silver nitrate $AgNO_3$ and hydrogen nitrate (nitric acid) HNO_3. The hydroxide radical OH is present in sodium hydroxide (caustic soda) NaOH. The ammonium radical NH_4 acts in place of a metal as in ammonium chloride NH_4Cl (used in Leclanché cells) and ammonium hydroxide NH_4OH (used in making explosives).

Composition of the Atmosphere

More than 99% of the atmosphere is composed of two gases, nitrogen and oxygen, in the proportion of about 4:1. Traces of about a dozen other gases do exist, the most important one of which is carbon dioxide. (Table 13.5.)

We have excluded such variable components as water vapour, which may account for anything up to 5% of the atmosphere, and localized pollution.

Table 13.5   Gases Present in the Atmosphere

Gas	Symbol	Per cent
Nitrogen	N_2	78
Oxygen	O_2	21
Argon	Ar	0.93
Carbon dioxide	CO_2	0.033
Neon	Ne	0.0018
Helium	He	0.00052
Methane	CH_4	0.0002
Krypton	Kr	0.0001
Hydrogen	H_2	0.00005
Nitrous oxide	N_2O	0.00005
Xenon	Xe	0.00001

The pollution may occur naturally or be man-made, and includes poisonous sulphur compounds, radioactive substances, ozone, carbon monoxide, and even solids in suspension, such as soot. Some diseases flourish in polluted areas, and frequently control is delayed because there is a long lapse of time between exposure to the pollution and the acute stage of the disease. This is the case with cigarette smoking which causes lung cancer, and also with radioactive exposure which causes anaemia.

The proportions of the most important atmospheric constituents—oxygen, nitrogen, and carbon dioxide—are maintained by cycles of interchange involving the earth's surface and the life it supports. (Figs. 13.3 and 13.4.) Most of the atmosphere exists within a height of 10 km above the earth. This region contains the clouds and the weather and is called the troposphere. (Fig. 13.5.)

Above this is the stratosphere in which jet planes but not piston engined planes can fly. High winds blow in the stratosphere which jets use to shorten their journeys. A band of ozone in the stratosphere helps to shield the earth from ultra-violet light. Even so the fraction which gets through can cause sunburn and snow blindness.

The ionosphere is higher still and is also a vital shield against excessive u-v waves. Cosmic rays which include streams of high energy particles from outer space are mainly absorbed by the ionosphere. Shorter wave radio waves bounce off the ionosphere and so reach distant parts of the world. Television waves are not reflected by the ionosphere so we have to beam them to relay satellites which we put into stable orbits at these altitudes.

Beyond the ionosphere is the exosphere which is true space where there are fewer molecules per cubic metre than any vacuum we can achieve on earth.

Source of the Atmosphere

We can assume, since they are the commonest gases in the universe, that hydrogen and helium were the original gases surrounding the earth. Being very light gases, however, they were lost into space in the early stages of the planet's life.

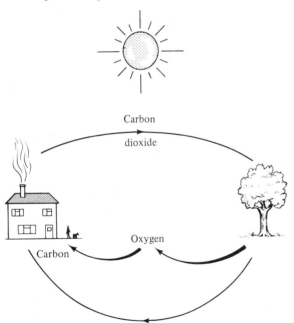

Fig. 13.3 The carbon cycle. Combustion, organic decay and respiration produce carbon dioxide from oxygen and carbon and release energy. Plants use the sun's energy to convert carbon dioxide into oxygen which is released, and carbon which provides material for growth

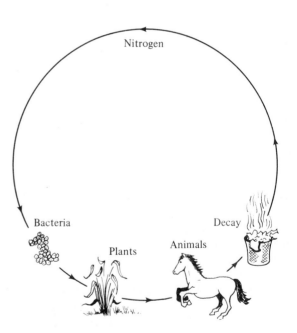

Fig. 13.4 The nitrogen cycle. (a) Bacteria extract free nitrogen. (b) Plants use this nitrogen from bacteria. (c) Animals eat the plants and gain size. (d) Dead plants and animals decay to release nitrogen.

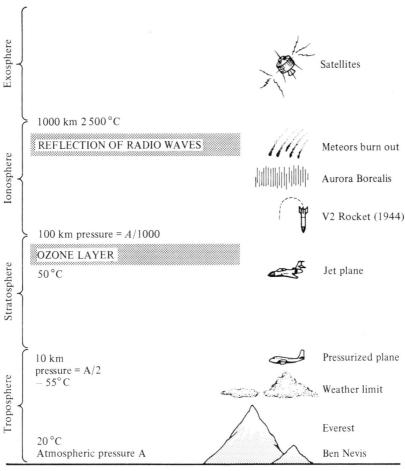

Fig. 13.5 The atmosphere above the earth

The present atmosphere has come from the earth's interior. Intense volcanic activity first produced stable compounds, such as nitrogen, carbon dioxide, and water vapour. Later, as the planet cooled, the water vapour condensed and life developed. The oxygen in the atmosphere was originally produced, and is now maintained, by the action of the plants which grow by absorbing carbon dioxide and releasing oxygen. The traces of the inert gases, argon, krypton, xenon, etc., in the atmosphere are a result of the radioactive decay or spontaneous fission of the elements present in the earth.

Oxidation

Fire is terrifying to animals but fascinating to man. Burning in air is a spectacular form of chemical reaction and one of the first that man studied. Observation of a fire suggests that matter is being lost in the smoke and flames. Magnesium tape when ignited burns rapidly giving off heat and smoke to become a white powder. When we weigh the powder, surprise, surprise, it weighs more than the original metal. The explanation is that the magnesium combines with oxygen from the atmosphere to form magnesium oxide.

$$2Mg + O_2 = 2MgO$$

$$\frac{2 \text{ molecules}}{\text{magnesium}} + \frac{\text{molecule}}{\text{oxygen}} = \frac{2 \text{ molecules}}{\text{magnesium oxide}}$$

During the combustion the heat, light and 'smoke' appear substantial but do not contain much matter. We can easily show that magnesium will not burn in a vacuum or when all the oxygen in the air has been used up. (Fig. 13.6.)

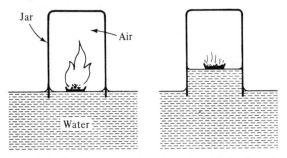

Fig. 13.6 Burning magnesium is extinguished when the oxygen in the air is used up

The formation of the compound magnesium oxide is an example of oxidation. It is a chemical reaction in that it takes place between molecules in fixed proportions and results in a compound normally unlike its constituents.

Copper foil heated in a flame develops a black tarnish which is due to oxidation at the surface.

$$2Cu + O_2 = 2CuO$$

Strongly heated zinc produces a yellow oxide which turns white on cooling.

$$2Zn + O_2 = 2ZnO$$

In each case oxygen from the air combines with the metal to form a metal oxide which contains more matter than was in the metal.

Some substances will combine with oxygen in more than one way to produce different oxides. We have seen that carbon burns to form the gas carbon dioxide.

$$C + O_2 = CO_2 \text{ (carbon dioxide)}$$

Pure carbon burns to nothing since carbon dioxide is gaseous and escapes. If the oxygen supply is restricted during the burning of carbon then a different oxide is produced which is of special interest because it is poisonous.

$$2C + O_2 = 2CO \text{ (carbon monoxide)}$$

Petrol is a compound of hydrogen and carbon and it burns incompletely in automobile engines. The exhaust contains carbon monoxide and is lethal when inhaled for a period.

Oxidation which occurs in air occurs much faster in pure oxygen. A mixture of hydrogen and oxygen burns to give water and is used in welding at a temperature of 2800°C. (Fig. 13.7.)

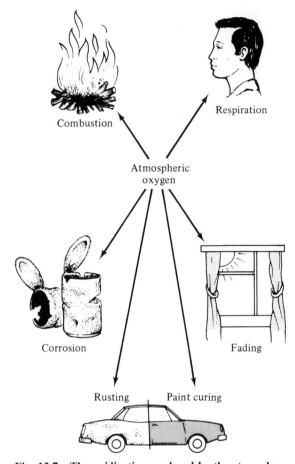

Fig. 13.7 The oxidization produced by the atmosphere

Oxyacetylene flames are even hotter at 3200°C. The storage and control of pure oxygen is a great hazard since the smallest particle of grit in the supply line can cause a spark which starts a self sustaining oxidation of the material of valves and gauges.

Corrosion

Oxidation can also take place at low temperatures when the process is very much slower. (Fig. 13.7.) In respiration carbon combines with the oxygen in the air to form carbon dioxide. The energy released by this process provides the energy which keeps plants and animals alive. Another slow oxidation process is the rusting of iron (symbol Fe). It is not simple however and the presence of water is essential to the process. Iron does not rust in dry air, nor in air free water. (Fig. 13.8.) It rusts in

142

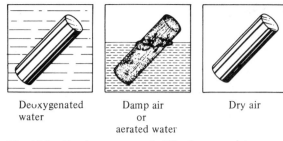

Deoxygenated water   Damp air or aerated water   Dry air

Fig. 13.8 Rusting occurs only if both water and oxygen are present

damp air or in water containing dissolved air. Rust has the chemical name of hydrated ferric oxide and an approximate formula of $2Fe_2O_3 3H_2O$.

Rust is granular and porous and so it does not protect the underlying metal against further attack. Once rusting has begun it is very difficult to protect the surface by painting. The rust may continue under the paint or the outer layer of rust may drop off with the paint. Zinc coating, called galvanizing protects steel very well as does chromium plating although neither can be applied in situ. Descaling followed by two-part epoxide paint works well. Alternatively the surface can be treated with dilute phosphoric acid containing phosphate. The surface is converted to a layer of crystalline metallic phosphate which provides a good base for painting. A grease or oil film will protect against rusting provided it is continually maintained.

Corrosion is big business. If we are not spending money on protection we are spending it on replacement of metal objects. The cost has been estimated to be as much as £1 per week for every person in an industrialized country. The main culprit is rust; 10% or more of the production of steel is to replace rusted parts. (Fig. 13.9.)

Not all corrosion is undesirable. Copper exposed to the atmosphere forms an attractive green sulphate coating which is impervious so that it both decorates and protects. Copper roofs are often used as an architectural feature. Aluminium forms a film of oxide on its surface which is protective. The film is so thin that it is invisible and does not impair the reflective properties of the surface. Aluminium mirrors are used in optical systems because they keep their shine. Electrolytic effects caused by impurities in the metal or the presence of other metals can increase corrosion severalfold. See page 152. Equally the presence of atmospheric pollution also affects corrosion, e.g., chlorine and sulphur. Hundreds of years ago silver objects remained bright as gold still does. The build up of sulphur pollution in the atmosphere now tarnishes silver and frequent polishing is necessary to keep the shine.

Extraction of Metals (reduction)

The removal of oxygen from a compound is called 'reduction'. Metals are extracted from their ores

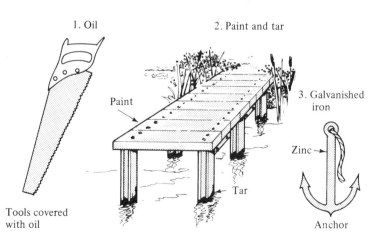

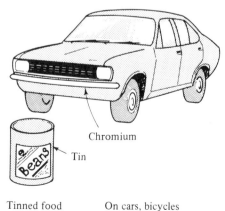

1. Oil

2. Paint and tar

Paint

Tar

Tools covered with oil

3. Galvanished iron

Zinc→

Anchor

4. Tinplate and chromium plating

Chromium

Tin

Tinned food

On cars, bicycles

Fig. 13.9 Preventing rust

by this process. The metal ore is usually heated together with a reducing agent such as carbon which has a greater attraction for the oxygen than the metal. For example tin is obtained by heating its ore with coke.

$$\text{Ore} + \text{Coke} \rightarrow \text{Tin} + \text{Carbon dioxide}$$
$$SnO_2 + 2C \rightarrow Sn + 2CO_2$$

Iron is extracted by a similar process.

$$\text{Iron ore} + \text{Coke} \rightarrow \text{Iron} + \text{Carbon monoxide}$$
$$Fe_2O_3 + 3C \rightarrow 2Fe + 3CO$$

The process is carried out in a blast furnace and is rather more complex than this simple equation suggests.

Limestone is added to mix with the impurities to form a molten slag which can be tapped off. (Fig. 13.10.)

Zinc is extracted from its ore zinc sulphide by first heating it to form zinc oxide, then it is mixed with powdered coal and heated to reduce it to the metal.

$$\text{Zinc oxide} + \text{Carbon} = \text{Carbon monoxide} + \text{Zinc}$$
$$ZnO + C \rightarrow CO + Zn$$

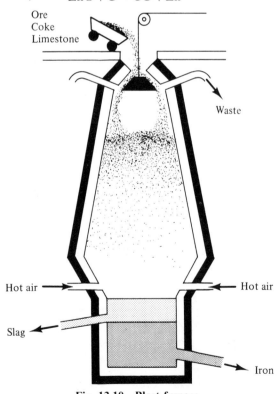

Fig. 13.10   Blast furnace

Acids

Sulphuric acid H_2SO_4 is a combination of two hydrogen atoms and a sulphate radical. The structure of the sulphate is such that it traps two electrons belonging to the hydrogen atoms. This makes the sulphate negatively charged and the hydrogen positively charged and the attraction of the opposite charges holds the molecule together. (Fig. 13.11.)

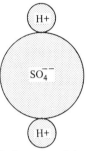

Fig. 13.11   The hydrogen sulphate molecule holds together by electrostatic attraction

When sulphuric acid dissolves in water the forces of attraction are reduced. A large proportion of the molecules dissociate into hydrogen ions and sulphate ions.

$$H_2SO_4 \rightarrow 2H^+ + SO_4^-$$

The semi-detached state of the hydrogen ions gives the acid its special reactive qualities.

All acids consist of hydrogen chemically combined with another element or radical. What makes them acids is the fact that they produce hydrogen ions when they are in solution.

Hydrochloric acid $HCl \rightarrow H^+ + Cl^-$
Nitric acid $HNO_3 \rightarrow H^+ + NO_3^-$
Carbonic acid $HCO_3 \rightarrow H^+ + CO_3^-$

The strength of an acid solution, its acidity, is determined by the density of hydrogen ions.

Acids are nasty substances. In concentrated form they burn clothes and flesh and dissolve metals. In dilute form they speed up the corrosion of metals and kill bacteria and have a sharp acidic taste. Fruit is acidic, e.g., lemons (citric acid) and so is vinegar (acetic acid).

Alkalis and Bases

When some substances dissolve in water they form

an electrolyte by producing hydroxyl ions (OH^-). These substances are called alkalis, e.g., sodium hydroxide (caustic soda).

$$NaOH \rightarrow Na^+ + OH^-$$

potassium hydroxide (caustic potash)

$$KOH \rightarrow K^+ + OH^-$$

These ions carry electric charge between the electrodes and thus a current passes. The OH (hydroxyl) combination is a radical, i.e., it acts as an entity. The strength of the alkali solution is determined by the density of the OH^+ ions.

Alkaline solutions feel soapy to the touch. When in strong solutions alkalis dissolve grease and attack the skin, e.g., ammonia NH_3. They are poisonous and dangerous to eyes. The colour of certain vegetable dyes are changed by alkalis. Litmus changes from red to blue in an alkaline solution. Thus paper impregnated with Litmus is used as an alkalinity indicator. The general name for any substance which has the hydroxyl radical OH is a Base. An alkali is the special case of a base which is soluble in water.

Neutralization of Acids

Water is one of the commonest substances on earth and one reason for this is the great affinity of hydrogen and oxygen. Given the right conditions they go for each other in a big way.

Having said that, let us look at what happens when an acid and an alkali are brought together. The acid provides H^+ ions and the alkali provides OH^- ions. The hydrogen and hydroxyl ions combine energetically to give water. The metal and the sulphate form a salt (potassium sulphate).

Hydrochloric acid + Sodium hydroxide

$$HCl + NaOH \rightarrow H^+ + Cl^- + Na^+ + OH^-$$
$$= NaCl + H_2O$$
$$= \text{Sodium chloride salt} + \text{water}$$

i.e.,

Acid + Base = Salt + Water

The acid is neutralized by the base to give salt and water. This statement is a general description of a whole range of reactions in which an acid is neutralized, e.g.,

Sulphuric acid + Caustic soda
$$= \text{Sodium sulphate salt} + \text{Water}$$
$$H_2SO_4 + NaOH = NaSO_4 + H_2O$$

In this type of reaction the acid is neutralized by the base, i.e., the concentration of H^+ ions in the acid is reduced by the concentration of OH^- ions in the base. The resulting product is acidic or basic according to whether the H^+ or OH^- ions predominate.

pH Values

The acidity–alkalinity of a solution can be represented on a scale from 1–14 described as 'potency of hydrogen' or pH values.

Strong acids have a pH of 1 while strong alkalis have a pH of 14. Water is neutral because it has H^+ ions and OH^- ions in equal proportions. Water and other neutral solutions have a pH of 7.

pH values are measured accurately by an electrical instrument, the pH meter. Alternatively chemical indicators can be used. The simplest is litmus paper which is red in acid solutions and blue in alkaline solutions. Other standard indicators change colour at certain pH values.

pH values are important sometimes even critical in many industrial and biological processes. Human blood is slightly alkaline with a pH of 7.4 while gastric juices are acidic and can have a pH value of 1.5. Values of soil pH are important to

Table 13.6   Typical pH Values of Some Substances

Properties	pH	Example of materials
Strongly acidic	1	Concentrated sulphuric acid
	2	Lemon
	3	Vinegar
	4	
	5	Bananas
Weakly acidic	6	Milk
Neutral	7	Pure water
Weakly alkaline	8	
	9	
	10	Tooth paste
	11	Antacid tablets
	12	Washing soda solution
	13	Lime water
Strongly alkaline	14	Caustic soda solution

farmers and gardeners. Geraniums like alkaline soil and grow well near to old brick walls, whereas azaleas like the acid conditions to be found where leaves collect and decay. The brewing of beer and the production of wine are highly sensitive to pH values.

Questions

1. Why do we use iron so widely even though it corrodes so readily?

2. Why would you expect strongly exothermic reactions to be stable?

3. Which came first, plants or animals?

4. If you draw a circle to represent the earth can you represent the highest clouds on the same scale?

5. Aluminium and copper corrode rapidly so why do they last so long exposed to the atmosphere?

Multiple Choice

1. Water is represented by the symbol
   (a) OH, (b) HO_2, (c) H_2O, (d) O_2H.

2. The smallest part of an element that can take part in a chemical change is
   (a) an atom, (b) a molecule, (c) a radical, (d) a compound.

3. Which molecule is triatomic?
   (a) ozone, (b) nitrogen, (c) hydrogen, (d) chlorine.

4. Acid + base gives a salt and
   (a) oxygen, (b) hydrogen, (c) water, (d) chlorine.

5. The ratio of nitrogen to oxygen in the atmosphere is approximately
   (a) 2:1, (b) 4:1, (c) 20:1, (d) 100:1.

6. The most common gas in the atmosphere is
   (a) oxygen, (b) hydrogen, (c) nitrogen, (d) argon.

7. Which gas is not important for living matter?
   (a) oxygen, (b) nitrogen, (c) argon, (d) carbon dioxide.

8. Which is *not* an inert gas?
   (a) argon, (b) neon, (c) chlorine, (d) xenon.

9. Which is the acid?
   (a) HCl, (b) H_2O, (c) NaCl, (d) CO_2.

10. Which is the base?
    (a) HCl, (b) KOH, (c) H_2SO_4, (d) NaCl.

11. Which completes the equation?
    $2Mg + O_2 =$
    (a) MgO, (b) MgO_2, (c) $2MgO_2$, (d) 2MgO.

12. A compound formed from a mixture of its components usually differs from the mixture in
    (a) density, (b) appearance, (c) mass, (d) properties.

13. How might you separate substances which are chemically combined?
    (a) magnet, (b) sieve, (c) chemical reaction, (d) flotation.

14. The symbol O_2 represents
    (a) two separate atoms, (b) a monatomic molecule, (c) a diatomic molecule, (d) a triatomic molecule.

15. About how many different elements are there?
    (a) 3, (b) 50, (c) 100, (d) more than 1000.

16. About how many different types of molecule are there?
    (a) 50, (b) 100, (c) 1000, (d) more than 1000.

17. What is the total number of atoms in one molecule of sulphuric acid? Formula H_2SO_4
    (a) 3, (b) 5, (c) 7, (d) 9.

18. Most of the atmosphere is contained within what distance of the earth?
    (a) 10 km, (b) 100 km, (c) 1000 km, (d) 10 000 000 km.

19. All acids contain
    (a) oxygen, (b) sulphur, (c) chlorine, (d) hydrogen.

20. Carbon dioxide is one compound of oxygen and carbon, the other is represented by the symbol
    (a) C_2O, (b) CO_2, (c) CO, (d) CO_3.

Multiple Choice Answers

1(c), 2(a), 3(a), 4(c), 5(b), 6(c), 7(c), 8(c), 9(a), 10(b), 11(d), 12(abd), 13(c), 14(c), 15(c), 16(d), 17(c), 18(a), 19(d), 20(c).

14.
Electrolysis

An electrolyte is a substance which in solution or in the molten state conducts electricity and is decomposed by it. Electrolytes generally conduct electricity well—not so well as metals but much much better than insulators. (Fig. 14.1.)

Now for conduction to occur charge carriers must exist and they must be free to move. In metals the charge carriers are electrons and we may well ask what form the carriers take in electrolytes. When certain compounds dissolve in water the forces holding the parts of the molecule together are reduced. Some of the molecules separate into parts called ions which are electrically charged, for example hydrochloric acid which is a compound of hydrogen and chlorine symbol HCl dissociates into hydrogen and chlorine ions

$$HCl \rightarrow H^+ + Cl^-$$

The chlorine ion keeps hold of a negative electron which is normally part of the hydrogen atom. This gives the chlorine ion a negative charge, and leaves the hydrogen ion positively charged. Even in a dilute solution there are millions of ions in each drop but the liquid as a whole is neutral since the positive and negative ions exist in equal numbers. If electrodes are placed in the solution and an electrical field applied the ions drift towards the electrode of opposite sign. Using carbon electrodes in a concentrated solution, the flow of current liberates hydrogen and chlorine gas. (Fig. 14.2.)

Good or 'strong' electrolytes are formed from certain salts or acids which dissociate almost completely into ions when they dissolve in water. Examples of strong electrolytes are common salt (sodium chloride) copper sulphate and sulphuric acid.

Substances which dissociate very little on dissolving are said to be 'weak' electrolytes. They are not such good conductors because they have fewer ions to carry the charge. Examples of weak electrolytes are acetic acid (vinegar) ammonium hydroxide and water.

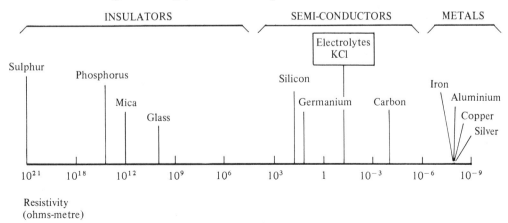

Fig. 14.1 Electrolytes have resistances nearer to metals than insulators (but note the powers of 10 scale)

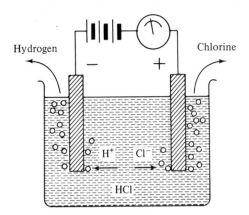

Fig. 14.2 Electrolysis of hydrochloric acid solution

Non-electrolytes do not dissociate on dissolving and have high resistances, e.g., alcohol, sugar. The equipment used to pass a current through an electrolyte is called a voltameter. For historical reasons we are stuck with that name but it is a confusing one since it is not a meter and it does not measure volts!

Electrolysis of Water

Pure water is a weak electrolyte and only dissociates into ions to a small degree which we can represent by.

$$H_2O \rightleftharpoons \underset{\text{hydroxyl ion}}{OH^-} + \underset{\text{hydrogen ion}}{H^+}$$

Because the number of molecules which dissociate is so small, pure water is an insulator. Very impure water conducts electricity well enough to have caused many fatal accidents to electricians working in damp conditions. (Fig. 14.3.)

A very small amount of impurity, for example hydrogen sulphate (sulphuric acid H_2SO_4) makes water into a much better conductor. The hydrogen sulphate is a strong electrolyte and dissociates readily as follows:

$$H_2SO_4 \rightleftharpoons 2H^+ + SO_4^{--}$$

The hydrogen and sulphate ions carry the charge between the electrodes. (Fig. 14.4.) When the charge arrives at the electrode however it is the water which is split up and its components released.

Fig. 14.3 Damp + electricity = danger

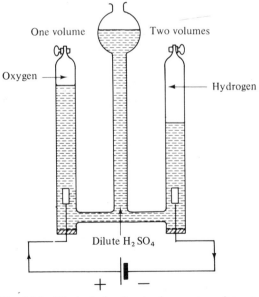

Fig. 14.4 Electrolysis of water in a water voltameter

148

At the Anode The hydroxyl ions combine to give water and oxygen

$$OH^- + OH^- = H_2O + O$$

Oxygen is released and the negative charge given to the anode.

At the Cathode Hydrogen ions combine in pairs and acquire electrons to give hydrogen molecules.

$$H^+ + H^+ + 2 \text{ electrons} = H_2$$

Hydrogen is released and a negative charge is taken from the cathode. Hence the result is that a current flows and the water is split into hydrogen and oxygen in the ratio of 2:1 by volume. (Fig. 14.4.) (Never mix these gases in these proportions—they are explosive.) Note that the sulphate ion takes part in the conduction but it is not released at the anode. The result is that the acid solution gets stronger near to the anode. This will disperse slowly by diffusion or more quickly if stirred.

The decomposing of water into hydrogen and oxygen has been suggested as a major use of off peak electricity. Electrical energy from natural sources like solar panels, wind vanes or hydro-electric plants often exceeds demand. The hydrogen produced from the electricity could be used as a fuel to increase capacity or reduce demand at peak times.

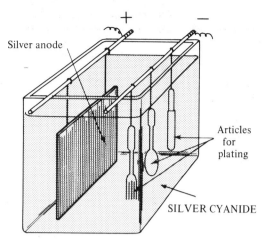

Fig. 14.5 Electroplating

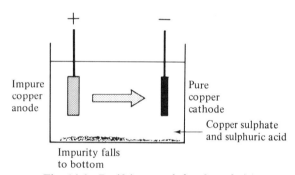

Fig. 14.6 Purifying metals by electrolysis

Electroplating

Electrolysis is used for many industrial processes. Silver or gold can be deposited as a thin layer onto more corrodable metals. The object to be silver plated acts as the cathode dipping into an electrolyte of silver cyanide. (Fig. 14.5.) The current must be low and the rate of deposition slow, to give a hard coating. Washing and burnishing gives the object a bright metallic finish.

Some metals, e.g., copper can be refined by electrolysis. The impure metal is made the anode and a suitable dissolved salt of the metal is the electrolyte. The passage of a current dissolves the anode and deposits pure metal on the cathode while impurities go into solution or fall to the bottom. (Fig. 14.6.)

Aluminium is deposited by electrolysis from an electrolyte made of its molten ores. These processes require enormous currents of the order of 100 000 amperes and so they are usually carried out near sources of cheap electrical energy, i.e., hydroelectric power stations.

$$\text{Mass deposited} = EIt \text{ (kg)}$$
$$\text{where } I = \text{current (A)}$$
$$t = \text{time (s)}$$
$$E = \text{a constant for the substance}$$
deposited called the electro-chemical equivalent
(Table 14.1)

The equation can be used to calibrate an ammeter to a high degree of accuracy since it is easy to measure accurately the time and the mass deposited.

Table 14.1 Electrochemical Equivalents
(mass liberated by 1 ampere in 1 second)

Element	Electrochemical equivalent (kg/A s)	
Aluminium	93.6	10^{-9}
Chlorine	367.6	10^{-9}
Copper	329.5	10^{-9}
Hydrogen	10.45	10^{-9}
Oxygen	82.9	10^{-9}
Potassium	405.3	10^{-9}
Silver	1118.3	10^{-9}
Sodium	238.4	10^{-9}
Zinc	338.8	10^{-9}

Problem

A current of 16 A flows through a silver voltameter for 30 minutes. Calculate the mass of silver deposited (electrochemical equivalent of silver $= 1118 \times 10^{-9}$ kg/A s).

Solution

$$M = EIt$$
$$= 1118 \times 10^{-9} \times 16 \times 30 \times 60 \text{ kg}$$
$$= 0.032 \text{ kg}$$
$$= 32 \text{ g} = \text{mass of silver deposited}$$

Problem

An ammeter was placed in series with a silver voltameter and recorded a steady current of 1.4 A. After 50 minutes 5.03 g of silver was deposited. What is the error in the ammeter?

Solution

$$M = EIt$$

Therefore

$$I = \frac{M}{Et} = \frac{5.03}{1118 \times 10^{-9}} \times 50 \times 60$$

$$= 1.5 \text{ A}$$
Error in meter $= 1.5 - 1.4 = 0.1$ A low

Electrode Potentials

When a metal such as zinc is placed in a dilute acid metal ions Zn^+ go into solution leaving the electrode negatively charged. (Fig. 14.7.) A double

0.76 Volts (for a standard electrolyte)

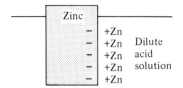

Fig. 14.7 Zinc ions go into solution and set up a potential difference

layer of charge builds up and eventually prevents any further net transfer of ions. This sets up an electrical potential difference between the electrode and the liquid. Different metals exert different electrode potentials and we can arrange them in a definite order according to their electrode potentials. (Table 14.2.) This arrangement of the metals

Table 14.2 Electrochemical Series

Element	Electrode potential in volts
Lithium	−3.02
Potassium	−2.92
Sodium	−2.71
Magnesium	−2.34
Aluminium	−1.66
Zinc	−0.76
Iron	−0.44
Cobalt	−0.29
Nickel	−0.23
Tin	−0.14
Lead	−0.12
Hydrogen	0.00
Copper	+0.34
Mercury	+0.80
Silver	+0.80
Gold	+1.68

is fundamental and is known as the electrochemical series. It is very useful in explaining and predicting the electrical and chemical properties of metals. The position of the metals in the series is the all important consideration rather than the actual values of the potentials.

In fact measuring the contact potential of one metal is difficult since the electrical measurement requires two contacts. The second electrode of course exerts its own electrode potential which influences the measurement. We solve the problem by measuring all the potentials relative to one substance, hydrogen. It is easier said than done

150

but a hydrogen electrode is obtained by bubbling the gas over a platinum electrode coated with platinum black. On this basis hydrogen itself has zero potential and relative to it other metals have either a positive or negative electrode potential. (Table 14.2.)

Since the potential is dependent on the dilution of the electrolyte, they are measured for each metal relative to a standard solution of one of its salts.

Primary Cells

We have seen that a metal placed in an acid develops an electrode potential. Put in another metal and it also exerts an electrode potential. There is a potential difference between the two electrodes which depends on the difference of their electrode potentials. The difference increases with the distance apart of the metals in the electro-chemical series.

Take the case of two widely separated metals zinc and copper in sulphuric acid. The difference in electrode potentials is:

$$+0.34 - (-0.76) = 0.34 + 0.76 = 1.1 \text{ v}$$

when the electrodes are connected a current flows which is due to the conversion of chemical energy into electrical energy. The zinc goes into solution and is gradually dissolved while hydrogen is released at the copper electrode. (Fig. 14.8.)

$$Zn^+ + H_2SO_4 = Z_nSO_4 + H_2^+$$

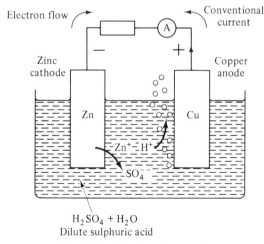

Fig. 14.8 Simple primary cell

Electrically what happens is this. The zinc ions dissolve and leave behind a surplus of electrons at the cathode. The hydrogen ions draw electrons from the anode to be neutralized and released as a gas. In the outside circuit electrons flow from cathode to anode, i.e., conventional current flows from the copper to the zinc. Unfortunately this simple type of cell is not much use as a source of electric current. The hydrogen bubbles quickly cover the copper which both reduces the voltage and introduces a high resistance into the circuit. The current reduces almost to zero and the cell is said to 'polarize'.

The Leclanché Cell

In another primary cell, the Leclanché, we overcome the problem of polarization by oxidizing the hydrogen as it forms at the anode. Carbon forms the anode, ammonium chloride is the electrolyte and manganese dioxide is the oxidizing agent. (Fig. 14.9.)

Powdered carbon which is a good conductor is mixed with the manganese dioxide to make the cell pass current more easily. The ammonium chloride is mixed with inert powder to form a paste. (Fig. 14.9.) The main applications of the cell is in 'dry' batteries used for radios and torches.

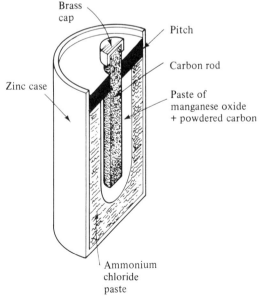

Fig. 14.9 Dry Leclanché cell

Mercury Cells

For the most miniature applications such as watches or cameras we use another primary cell, the mercury cell. It has a high energy-to-weight ratio and is used where a few milliamps of current are required. Typically a cell of the size shown in Fig. 14.10 has a capacity of about 1 ampere-hour and gives a voltage of just over 1 volt.

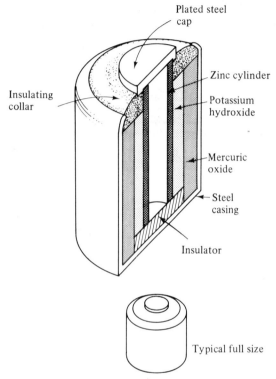

Fig. 14.10   Mercury cell

Chemical Reactivity of Metals

Metals high in the electrochemical series are very reactive. Potassium for example ignites and oxydizes spontaneously when exposed to air. Magnesium burns vigorously if heated. The tendency to oxydize reduces as we go down the series. Zinc oxydizes more readily than tin and tin more readily than lead. Below hydrogen come even less reactive metals starting with copper and ending with gold which is so unreactive it is found naturally as a metal not reacted with any other elements.

Metals higher in the series will displace those that are lower from solutions of their salts. For example zinc placed in silver nitrate solution dissolves to become zinc nitrate pushing out the silver which is precipitated. Dilute acids attack metals above hydrogen because the metal is dissolved and hydrogen is released. The further away the metal is from hydrogen in the series the more violent is the attack.

Electrolytic Corrosion

The position of a metal in the series casts light on its tendency to corrode. When several metals are in contact and connected by an electrolyte, like seawater or rain water, the metal highest in the series tends to corrode the most and its presence reduces the corrosion of the others. For this reason a zinc anode is often attached below the waterline of an iron ship. It is gradually corroded away in preference to the iron of the hull and is replaced from time to time. For the same reason copper fittings must be avoided on steel hulls or rapid corrosion of the steel occurs. The same considerations apply to brackets, pipe clips, rivets and valves on metal installations.

Impurities in metals can accelerate corrosion tremendously by setting up local action. Any particle of more electropositive metal impurity embedded in the base metal forms a primary cell. A local circulating current flows and the base metal is rapidly eroded. Pure zinc resists acid attack quite well whereas impure zinc placed in dilute sulphuric acid bubbles vigorously as it is dissolved.

The greater chemical reactivity of metals higher up the series is sometimes masked by other effects. Zinc and aluminium oxydize more readily than iron but their oxides form a protective layer which prevents further oxydations. Iron oxydizes more slowly, but rust is not impervious and gives no protection so that the process goes on and on eventually destroying the iron object.

Secondary Cells

In primary cells there is a one way change of chemical energy into electrical energy and the components are used up. In secondary cells the chemical changes which produce the current are

	Metal	With water	With common acids	With air
Base metals	Lithium Potassium Sodium	Reacts even when cold		Easily inflammable
	Magnesium Aluminium Zinc	Reacts at red heat	Attacked by dilute acids	Burns if heated
	Iron Nickel Tin Lead Hydrogen Copper	No action below white heat		Forms the oxide on heating but does not ignite
			Attacked by nitric acid	
Noble metals	Mercury Silver Gold	No action	No action	No action

reversed if a current is driven through the cell in the opposite direction. Thus secondary cells can be recharged again and again. They have a low internal resistance and do not polarize even when large currents are taken for a long time.

Lead Acid Accumulators

Lead acid accumulators are the most widely used storage cells. They provide a compact high capacity source of electrical energy for traction vehicles, car batteries and portable lighting. (Fig. 14.11.) The cell has a positive electrode (anode) of lead oxide and a negative electrode (cathode) of spongy lead. Ions of sulphuric acid H^+ and So_4^- carry the current between the electrodes.

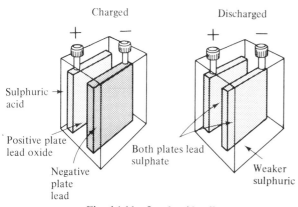

Fig. 14.11   Lead acid cell

On Discharge   Both plates turn to lead sulphate and the acid becomes diluted with water.

$$\left.\begin{array}{l}\text{Anode } PbO_2\\ \text{Cathode } Pb\end{array}\right\} + 2H_2SO_4 - 2PbSO_4 + 2H2O$$

On Charge   The electrodes are turned back to lead oxide and lead and the acid becomes stronger.

$$2PbSO_4 + 2H_2O \rightarrow 2H_2SO_4 + \left\{\begin{array}{l}PbO_2 \text{ at anode}\\ Pb \text{ at cathode}\end{array}\right.$$

During charge the relative density of the acid increases from 1.15 fully discharged to 1.25 fully charged. At the same time the potential difference rises from 1.8 V to 2.2 V.

Care of Lead Acid Batteries

The current capacity of a lead accumulator is measured in ampere hours. A car battery typically rated at 40 amphour will give 4 amperes for 10 hours or 8 amperes for 5 hours. If a battery is required to give a large current for example to activate the starter motor of a car, the plates must have a large surface area. To achieve this in a compact size the cells are made from as many as 9, 11 or 13 interleaved plates. Perforated plastic separators prevent electrical contact between the plates. The active substances are packed into holes

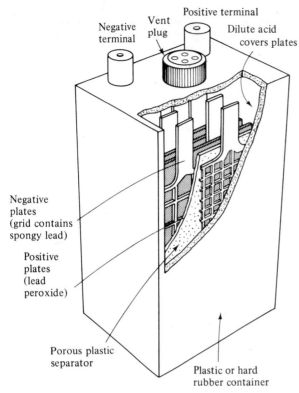

Fig. 14.12   Construction of a lead acid cell

Negative terminal

Vent plug

Positive terminal

Dilute acid covers plates

Negative plates (grid contains spongy lead)

Positive plates (lead peroxide)

Porous plastic separator

Plastic or hard rubber container

in the electrodes which are made of antimony lead alloy in the form of a grid. (Fig. 14.12.)

Batteries should not be allowed to become completely discharged nor should they be stored discharged. The lead sulphate crystallizes into a form that cannot be reconverted on charging. Charging should be slow, i.e., one or two amps. (Fig. 14.13.) Rapid charging makes the paste in the grids soft and liable to physical collapse. The hydrogen

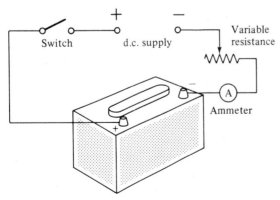

Fig. 14.13   Battery charging circuit

Switch

d.c. supply

Variable resistance

Ammeter

given off during the charge (gassing) must be free to escape and ventilation is necessary to prevent a dangerous build up of this inflammable gas. Do not check the acid level by the light of a match!

When too large a current is taken from a lead acid accumulator, paste drips from the plates which may buckle under the strain. Unless acid has spilled, only distilled water should be added to a battery to restore the water lost by gassing and by evaporation.

The life of a lead acid cell depends on its quality of construction and on the care it gets in use. It is a poor car battery that lasts less than two years and a good one may last up to five.

A fully charged 12 V battery of six cells will register about 14 V when a high resistance voltmeter is used for the measurement. To obtain a high voltage from a number of batteries we connected them in series. (Fig. 14.14(a).)

To obtain a large current capacity at low voltage the cells are connected in parallel. (Fig. 14.14(b).)

Measured with a high resistance voltmeter a worn out battery will give the same reading as a new one. It is only under load that the difference is apparent. Batteries are usually tested by measuring the voltage they give when under a standard load.

If the battery is small or if the effective plate area has been reduced due to old age or misuse then it will have a high internal resistance (see page 213). In this case any attempt to draw a heavy current will cause the voltage at the terminals to drop drastically. Even in a good battery this effect shows in the dimming of the lights or radio when the starter of a car is activated. With a poor battery the voltage available to the ignition coil may be insufficient to fire the plugs or in a bad case to turn the engine. It was a sad day when starting handles were discontinued by car makers!

Alkaline Cells

Alkaline cells cost more than lead acid cells but they store more electrical energy weight for weight, they are more mechanically robust and they stand up better to misuse and therefore last longer. They maintain their voltage better as they run down and they can deliver large currents or be left uncharged

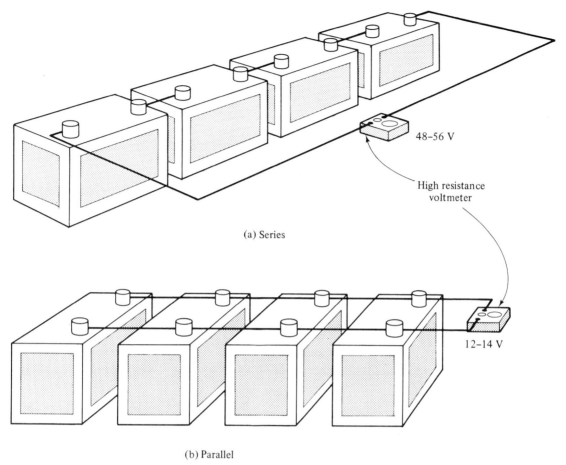

(a) Series

48–56 V

High resistance
voltmeter

(b) Parallel

12–14 V

Fig. 14.14   Use of a high resistance voltmeter to measure the voltage of cells in series and in parallel

for months without damage. They are suitable for yachts and railway rolling stock which may be unused for long periods.

The electrodes in the nickel–iron (Nife) cell have a grid structure which holds the active materials. (Fig. 14.15.) The electrolyte is potassium hydroxide (caustic potash).

Another form of alkaline cell is the nickel cadmium (nicad) cell which can supply larger currents and operate at lower temperatures than the Nife cell. It is similar in construction except that spongy cadmium replaces the finely divided iron. (Fig. 14.16.)

Alkaline cells in miniature form are used as rechargeable power sources for electronic calculators.

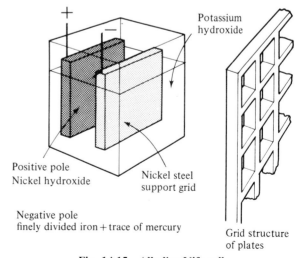

Potassium
hydroxide

Positive pole
Nickel hydroxide

Nickel steel
support grid

Negative pole
finely divided iron + trace of mercury

Grid structure
of plates

Fig. 14.15   Alkaline Nife cell

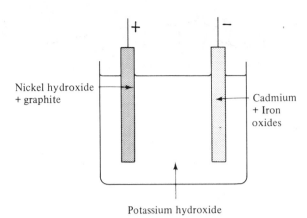

Nickel hydroxide + graphite

Cadmium + Iron oxides

Potassium hydroxide

Fig. 14.16   Alkaline nicad cell

Questions

1.   How is it that pure water and dry salt are insulators but salt solution is a conductor?

2.   What are the advantages and disadvantages of electroplating when compared with painting?

3.   How could the discovery of a super efficient lightweight battery revolutionize transport energy and pollution?

4.   Why would you expect a brass nut on a steel bolt to have a short life in a place exposed to moisture?

5.   Why have a large battery when a small one has the same e.m.f.?

Multiple Choice

1.   Which are the strong electrolytes?
(a) water, (b) common salt solution,
(c) carbon dioxide, (d) hydrochloric acid.

2.   What is released at the anode when a current is passed through platinum electrodes placed in dilute sulphuric acid?
(a) oxygen, (b) hydrogen, (c) sulphur, (d) water.

3.   When water is electrolyzed what is the ratio by volume of hydrogen to oxygen released?
(a) 3:1, (b) 2:1, (c) 1:1, (d) 1:2.

4.   Which metal is the least chemically reactive?
(a) zinc, (b) iron, (c) copper, (d) silver.

5.   Which is the secondary cell?
(a) Leclanché, (b) lead acid, (c) mercury,
(d) zinc–copper.

6.   Which is the more chemically reactive metal?
(a) iron, (b) sodium, (c) tin, (d) lead.

7.   Which is the most inert metal?
(a) gold, (b) aluminium, (c) magnesium, (d) zinc.

8.   Which two gases are given off when a lead acid battery is charged?
(a) oxygen, (b) chlorine, (c) hydrogen, (d) nitrogen.

9.   How many cells are there in a lead acid battery rated at 12 V.
(a) 12, (b) 6, (c) 4, (d) 1.

10.   The energy stored by a battery is rated in
(a) volts, (b) current, (c) amp-hours, (d) watts.

Problems

1.   State what is meant by an electrolyte and give an example.

2.   State the difference between a primary cell and a secondary cell and describe the action of one type of primary cell.

3.   Draw a labelled diagram of a lead acid cell indicating its main features.

4.   What is the effect of passing a direct current through water which contains a trace of sulphuric acid using platinum electrodes?

5.   You are provided with six 2 volt cells. How would you connect them up to give a battery of (a) maximum e.m.f., (b) minimum internal resistance? What would be the e.m.f. in each case? [(a) series 12 V; (b) parallel 2 V]

6.   Copper is refined by electrolysis using the impure metal as an electrode. If 200 A flow for 6 hours what mass of metal will be deposited at the cathode? Will the anode lose more or less mass than this? (Electrochemical equivalent of copper = 0.329 $\times 10^{-6}$ kg/C.) [1.42 kg, more]

7.   A current of 10 A flows through a silver voltmeter for 1 hour. What mass of silver is deposited on the electrode? (Electrochemical equivalent of silver = 1.12 $\times 10^{-6}$ kg/C.) [40.3 g]

Multiple Choice Answers

1(bd), **2**(a), **3**(b), **4**(d), **5**(b), **6**(b), **7**(a), **8**(ac), **9**(b), **10**(c).

15.
Vibrations and Waves

Simple Harmonic Motion (S.H.M.)

When a force is applied to a fixed particle it distorts the nearby matter to which it is attached.

When the applied force is removed, the environment pushes the particle back towards its original position. In returning, the particle gathers speed until, when it reaches the equilibrium position, it overshoots on the other side. (Fig. 15.1.) This action is repeated back and forth giving rise to a common event in nature—periodic motion. Such motion exactly repeats itself in equal periods of time T. The number of vibrations per second called the frequency f, equals the reciprocal of the period

$$f = \frac{1}{T}$$

For example a vibration of period $\frac{1}{5}$ second has a frequency of 5 hertz (1 hertz = 1 vibration per second). The greatest distance a particle moves from the equilibrium position is the amplitude a. Note that when the amplitude is increased the period remains the same. The object just moves quicker so that it covers the larger distance in the same time.

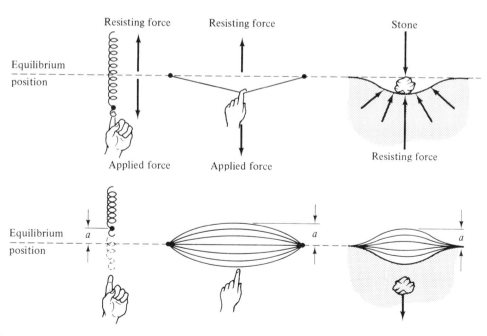

Fig. 15.1   An applied force produces a displacement which gives rise to a resisting force. When the force is removed, vibration occurs about the equilibrium position

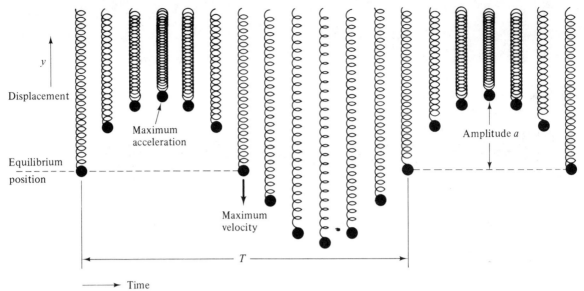

Fig. 15.2 **The position of the mass on the spring is shown at different instants as it vibrates with simple harmonic motion**

It often happens that the restoring force is directly proportional to the displacement of a particle from its equilibrium position. The resulting vibration is called simple harmonic motion or S.H.M. The mass suspended from the spring in Fig. 15.2 is an example of linear simple harmonic motion. The motion can also be angular as in Fig. 15.3.

Free, Forced, and Resonant Vibrations

A particle which is free to vibrate will assume a natural frequency of vibration which will depend on the mass of the particle and the elasticity of the medium.

If a different frequency is imposed on the par-

ticle, the amplitude of the particle will not be large, because the incident energy is dissipated as random vibration, i.e., heat.

Should the forced frequency coincide with the natural frequency, the particle increases its amplitude and stores the energy. This situation is called resonance. (Fig. 15.4.) Minor resonances can occur when the forced frequency, f, is a simple fraction or multiple of the natural frequency f_0.

$$f = f_0, \ 2f_0, \ 3f_0, \ \text{etc.}$$

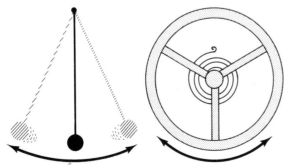

Fig. 15.3 **The pendulum bob and watch balance, oscillations of which are examples of angular simple harmonic motion**

Fig. 15.4 **Effect of the frequency of the forced vibration on amplitude**

Damping

There are two ways of driving a vehicle over an unmade or washboard road surface—either fast or very slow. At a certain intermediate speed the frequency of the road irregularities is near to the natural frequency of the springing, and the vibration builds up and makes control difficult. Even on apparently smooth roads, a recurring irregularity of the surface has been known to shatter car windscreens. This tendency to resonate can be reduced by providing a means of absorbing the undesirable energy of vibration. In motor vehicles, the energy is absorbed by hydraulic dampers (shock absorbers). (Fig. 15.5.) Movement in either

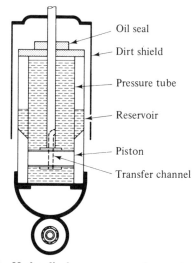

Fig. 15.5 Hydraulic dampers are used on motor vehicles to absorb the energy of the suspension. Limited resistance to vertical motion is provided as the liquid passes through the transfer channel

direction is resisted by the viscous forces as the liquid passes from one side of a piston to the other through a restricted aperture, called the transfer channel.

Instruments with moving pointers, such as balances and electrical meters, are damped to curtail oscillation about the final rest position.

It is important to provide just the right degree of damping. This is usually the damping which returns the system to the equilibrium position in the shortest possible time without oscillation. The condition is called *critical damping*. The graphs of Fig. 15.6 illustrate the effect of varying degrees of damping.

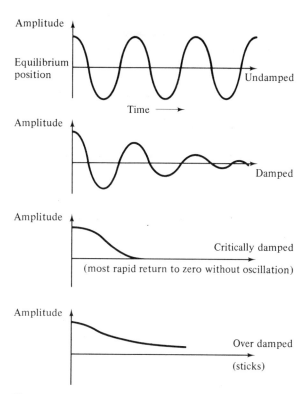

Fig. 15.6 The graphs show the behaviour of a particle which is displaced from its equilibrium position and released under varying degrees of damping

You can roughly check a car or motorcycle for critical damping by applying your weight to the suspension and then quickly removing it. The vehicle should return quickly to its original position without oscillating.

Energy Dissipation and Wave Motion

When energy is supplied at a point of disturbance the energy travels away from the point in the form of a pulse. If the pulse is confined to one direction, like a ripple along a piece of rope, each particle in its path will, in turn, assume the movement of the original particle. (Fig. 15.7.)

If the disturbance continues then a continuous periodic wave is generated. We can form a picture of such a wave by plotting the position, at one instant in time, of each particle in its path. (Fig. 15.8.) The distance between the wave crests or between corresponding points on the curve is the

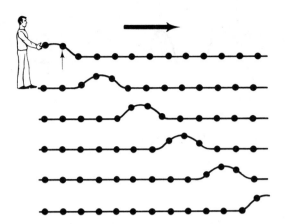

Fig. 15.7 A single pulse travels away from a source of energy. The pulse shown is only half a complete cycle and, in an elastic material, the amplitude remains constant

wavelength symbol λ. As the wave profile moves, it displaces the position of each particle on its path. All the particles on the wave vibrate but each one lags slightly behind the one on the left because of its inertia. Figure 15.9 shows two positions of a wave separated by a short time interval. The wave profile travels from left to right, but note that each particle moves up and down about its equilibrium position.

Speed of wave

$$= \frac{\text{distance travelled by wave}}{\text{time taken}}$$

$$= \frac{\text{number of waves} \times \text{wavelength}}{\text{time}}$$

i.e.,

$$\text{Speed} = \text{frequency} \times \text{wavelength}$$

$$v = f\lambda$$

Problem

One end of a rope is vibrated at 8 Hz and the waves generated measure 60 cm from crest to crest. How fast do they travel and what is the period of the vibration of any part of the rope?

Solution

$$v = f\lambda$$
$$v = 8 \times 0.60$$
$$\text{velocity} = 4.8 \text{ m/s}$$
$$\text{period} = \frac{1}{n} = \frac{1}{8} \text{ s}$$

Wave Behaviour

When energy of any kind involving a regularly fluctuating quantity travels along, it is called a wave. Wave motion is the most important means of energy transfer in the universe, whether it be light energy travelling from the sun to the planets or sound energy travelling from one end of a room to the other.

Waves in Strings

Plucking or vibrating a string generates a wave which travels along the string. (Fig. 15.8.) Particles of the string move transversely, i.e., at right angles

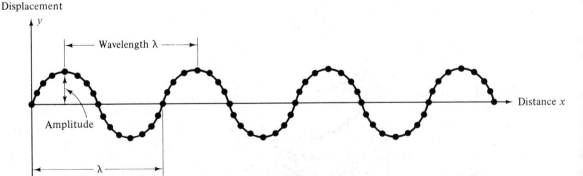

Fig. 15.8 A picture of a wave in a string frozen at an instant in time showing the displacement of particles along its path

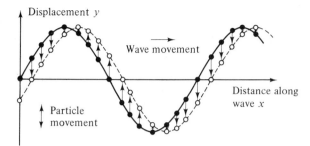

Fig. 15.9 Pictures taken of a wave at two close instants of time show that the particles do not move in the direction of the wave but at right angles to it

to the direction of the wave and hence a wave in a string is classified as a transverse wave.

The speed of a wave in a string varies with the tension and the mass per unit length as follows

$$\text{Speed} = \sqrt{\frac{\text{tension}}{\text{mass/unit length}}}$$

Compressional Waves

A mechanical compression wave can be transmitted as shown in Fig. 15.10. This is an example of a longitudinal wave. Notice that each link moves backwards and forwards in the direction of the wave.

Compressional waves travel invisibly through matter the molecules vibrating like the links in the spiral spring.

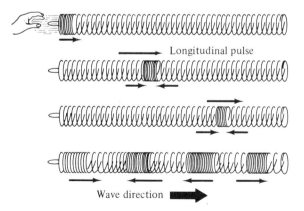

Fig. 15.10 A longitudinal pulse and a longitudinal wave are shown travelling along a spiral spring. Individual coils move back and forth about their undisturbed positions

Sound Waves

Let us first pose the question 'what is sound?' If we were to take suitable instruments between source and receiver while a sound wave is frozen in time, what would we find there? The instruments would show that the pressure along the wave changes regularly, alternating from a little above average to a little below average. (Fig. 15.11.)

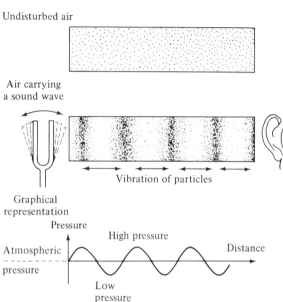

Fig. 15.11 In a sound wave the to and fro movements of the particles produce high and low pressure regions which move along as a wave

The compressions and rarefactions move through the material with the velocity of the wave. For a sound wave in air this is about 350 m/s, i.e., 1260 km/h.

The particles in the path of the wave move from their equilibrium position first in the direction of the wave and then in the reverse direction. This type of to and fro movement along the direction of travel makes sound a longitudinal wave.

Note that we can still represent a longitudinal wave by a graph.

Surface Liquid Waves

Waves in a liquid surface can be quite complex, like sea waves generated by a storm centre in the great oceans, or they can be more regular like the

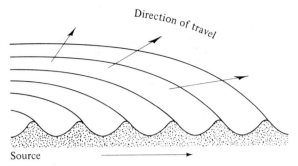

Fig. 15.12 **Liquid surface waves are transverse waves not having the same shape as the simple sine curve**

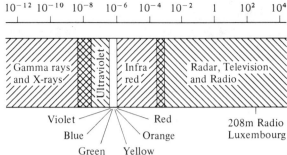

Fig. 15.14 **The electromagnetic spectrum. (Wavelengths shown in metres)**

waves on a still pond when a stone is dropped into it. (Fig. 15.12.) A wave on a liquid combines both longitudinal and transverse vibration which give each particle on its path a cyclic motion. The principal restoring force acting on particles in a sea wave is the force of gravity.

Electromagnetic Waves

Imagine a vibrating charge, which produces a changing electric field around it. This produces a changing magnetic field beyond that, which produces a changing electric field beyond that, and so on, one field following the other in leap-frog fashion. (Fig. 15.13.) In this way, energy travels away from the source as an electromagnetic wave. The electric and magnetic components are perpendicular to each other and to the direction of travel (transverse wave).

Since both electric and magnetic fields can exist in a vacuum, electromagnetic waves do not require any medium for their propagation and they can travel through empty space. The waves travel at the speed of light, 300 000 km/s in a vacuum, and nothing but nothing in the universe can travel faster! Electromagnetic waves include X-rays, ultraviolet rays, radar and radio as well as light

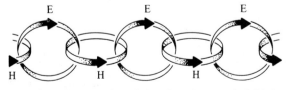

Fig. 15.13 **Propagation of electric and magnetic fields in an electromagnetic wave**

162

which differ from each other only in the length of the waves. (Fig. 15.14.)

Wave Intensity I

All waves are the means by which energy travels from a point where it is being generated. The intensity of a wave is the energy flowing per second per unit area perpendicular to the direction of the wave. Intensity is measured in watts per square metre.

If a wave diverges in all directions from a source, e.g., light from a light bulb, the intensity I reduces with the distance from the source according to an inverse square law.

$$I \propto \frac{1}{d^2}$$

In addition to this reduction the waves are reduced further by absorption by the medium, e.g., headlights absorbed by fog.

Reflection of Waves

When a wave meets a surface separating two materials in which its velocity is different, some of the energy of the wave is reflected. The reflected wave has the same velocity, frequency, and wavelength as the incident wave, and the two waves are symmetrical with the surface. (Fig. 15.15.)

Refraction of Waves

A change in medium or a change in conditions may cause a wave to slow down. This slowing-

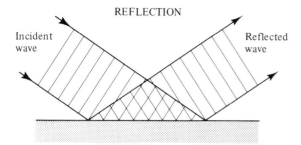

REFLECTION

Incident wave

Reflected wave

Fig. 15.15 The reflected wave and the incident wave are symmetrical with the surface

down has the effect of shortening the wavelength, that is, the distance between the crests of the waves. This is shown for normal incidence in Fig. 15.16 which also shows what happens when the wavefronts meet the surface obliquely. One end of each wavefront, the end which touches the obstacle first, slows up before the other, and the whole wavefront is made to swing round. The retardation of the waves causes deviation towards the normal at the surface. Conversely, if the waves travel faster in the second medium they deviate away from the normal. This phenomenon is called *refraction*. It may occur sharply as, for example, when light waves enter a glass block, or it may occur gradually as is the case for sea waves approaching a beach. The velocity of the waves reduces as the depth decreases, and the waves are bent around to face the shore. (Fig. 15.17.) The

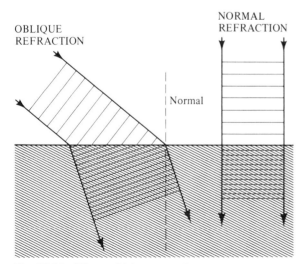

Fig. 15.16 Waves which are retarded at a surface reduce their wave-length and are refracted towards the normal

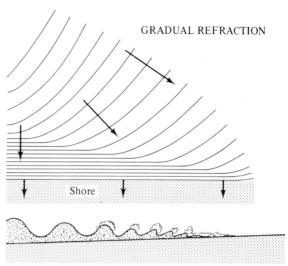

GRADUAL REFRACTION

Shore

Fig. 15.17 Sea waves retarded by a shelving beach swing round and break on the shore

shortened wavelength causes the waves to break as they become too steep-sided to support themselves.

Interference

Two wave trains will pass through the same region and each will emerge on the other side quite unaffected by the other. Where they actually overlap however, they do interact or interfere with each other.

We can appreciate the principle of interference by considering the interaction of single pulses travelling along a string of particles. Two crests travelling in opposite directions produce a maximum amplitude where they coincide and then move on unaffected. (Fig. 15.18.)

The peak explains the freak waves which occur at sea only rarely but occur frequently in the nightmares of small boat sailors.

Where a crest and a trough coincide, they cancel each other and result in a point of zero displacement. (Fig. 15.19.)

The complexity of the interference pattern from two sources of continuous waves may be appreciated from Fig. 15.20. The coincidence of two crests or two troughs produces a maximum amplitude or *antinode*. Where crest and trough coincide, they produce a minimum amplitude or *node*.

163

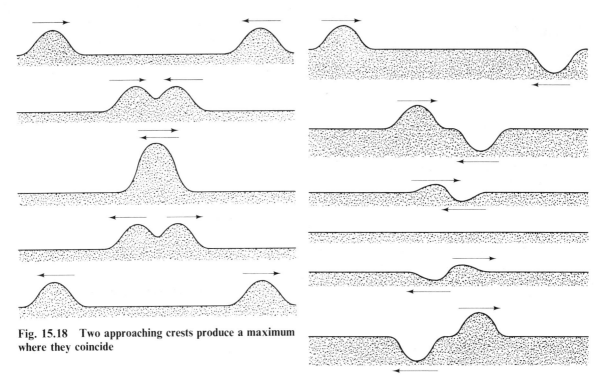

Fig. 15.18 Two approaching crests produce a maximum where they coincide

Fig. 15.19 The coincidence of a crest and a trough produces destructive interference

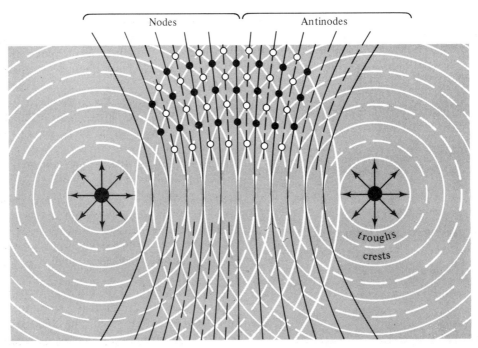

Fig. 15.20 As the waves of equal amplitude *and* frequency radiate from the sources, a pattern of lines of minimum amplitude (nodes) and lines of maximum amplitude (antinodes) is produced

Standing Waves

Directly between the sources of Fig. 15.20 the interference sets up a rather special type of wave which forms whenever two waves of equal frequency and amplitude travel in opposite directions through the same space.

At every stage in their motion, the waves sum to zero at certain fixed points, called nodes, situated half a wavelength apart. Because of these fixed nodes we call the resulting wave a standing or stationary wave.

Standing waves are very common because they are generated whenever a wave is reflected at right angles to a boundary. You may have seen one when you twang a stretched elastic band. The nodes are where the blurred image of the band is nipped in. The antinodes are the places where the amplitude is greatest. (Fig. 15.21.)

Fig. 15.21 Standing wave in an elastic band showing nodes and antinodes

When a standing wave is formed within the confines of an object we get the resonance described earlier in this chapter.

A rather dramatic example of transverse standing waves was the destruction of the Tacoma Suspension Bridge, USA, in 1940. The centre span of 2800 ft was slung between towers 420 ft high, which were anchored by cables to blocks containing 20 000 yd^3 of concrete. Even during construction it was found to have a tendency to vibrate and an effort was made to reduce it. When completed it resonated to a standing wave with eight or nine nodes at a frequency of 36 or 38 vibrations/minute. When it was completed, it lasted just four months and seven days, before one mildly windy day the amplitude of the standing wave rose to several feet, exceeding the elastic limit of the structural members, and the bridge broke up.

Shock Waves

Just as an object has a natural frequency of vibration, so in each medium a wave has a natural speed of propagation. It is in some cases possible for a body to move through the medium at a higher speed than waves. Bullets, shells, and projectiles travel much faster than sound and, now, aircraft can maintain supersonic velocities. Ships often exceed the speed of water waves but, in contrast, no object can ever travel faster than light waves in a vacuum.

When a body exceeds the speed of waves in a medium, no wave travels ahead of it to produce a mild disturbance. The air is undisturbed until the body itself arrives. The sudden disturbance which the object produces is called a *shock wave* and this travels away from the leading edge of the object at the wave velocity. (Fig. 15.22.) The bow wave of a

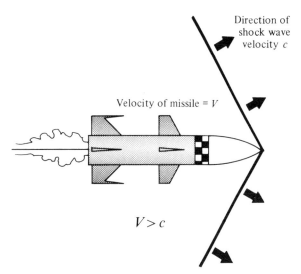

Fig. 15.22 Two positions of a missile and its shock wave

ship travelling faster than water waves is another example of a shock wave. We often express supersonic velocities in units of the speed of sound, called *mach numbers*.

$$\text{Mach number} = \frac{\text{speed of body}}{\text{speed of sound}}$$

Since the speed of sound varies with atmospheric conditions, so does the speed indicated by a certain mach number. Under normal conditions, a speed of mach 1 would correspond to about 1250 km/h and speeds of about mach 5 have been achieved by prototype 'planes, such as the American X15.

Properties of Electromagnetic Waves

When a beam of white light is passed through a prism it is dispersed into the colours of the rainbow. (Fig. 15.23.) The amount of refraction depends on the wavelength of light, i.e., the colour.

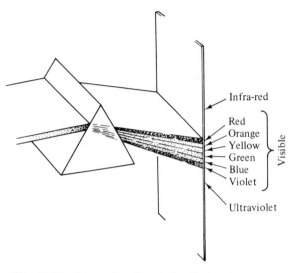

Fig. 15.23   Dispersion of optical radiations by a prism

The red light having longer wavelength is refracted through a smaller angle than the violet light. The colours contained in the white light are separated into the spectrum in the same order. Red, orange, yellow, green, blue, and violet.

The range of wavelengths we call light are only those electromagnetic radiations to which our eyes are sensitive.

Infra-red Radiation

The infra-red radiation has longer wavelength than red light. We cannot see it but we can feel it on our skins. It is heat radiation such as we get from electric fires and bathroom heaters.

Infra-red radiation penetrates fog and haze, and, by using suitable sensitive photographic plates, we can take clear photographs under conditions which would make ordinary photography impossible. Infra-red radiation may penetrate deeply into a substance before being absorbed and it can

thus be used to produce more even heating. This property finds an application in the quick drying of paints on cars and in the dehydration of foods.

Radio, TV, and Microwaves

Microwaves are next after infra-red in the length of their waves. Microwaves penetrate the body producing a mild degree of heat internally when they are absorbed. They are used medically to relieve arthritis and rheumatism. High intensity microwaves however produce intense internal heat and cook food from the inside in a matter of minutes. Be careful, they cannot distinguish between human hands and juicy steaks!

Following on microwaves in wavelength are radar, television and radio waves in that order. Radar, TV, and vhf radio are received by tuned electrical circuits (aerials) over line-of-sight distances. This restricts their range to about 50 miles even when operated from high buildings. Longer wavelength radio waves can reach over the horizon by being reflected from the ionosphere. Artificial satellites do an even better job by amplifying and retransmitting radio signals across large parts of the globe.

Ultra-Violet Rays

Below the violet end of the spectrum and of shorter wavelength is the ultra-violet radiation. This component of the sun's radiation makes plants grow. When it is absorbed by the skin it produces a tan in the short term and wrinkles in the long term.

Dangerous levels of ultra-violet radiation are to be found at high altitude and where the sun's rays are strongly reflected from snow or sand surfaces. In these settings, sun goggles or glasses must be worn for protection.

Ultra-violet radiation kills germs and is used in hospitals to sterilize instruments and room interiors. Food keeps longer in a fresh condition if it is irradiated with ultra-violet rays to kill the germs which cause decay.

X-rays

X-rays are produced by the rapid deceleration of

electrons as they strike the atoms of a substance. They are electromagnetic waves of such very short wavelength and so penetrating that at the time of their discovery scientists did not believe they could be electromagnetic, and hence gave them the name X for unknown rays.

By controlling the wavelength and intensity we can make the X-rays just penetrate the flesh of the body but not the bone. In this way the internal structure is revealed as a shadow on a photographic plate. Within weeks of the discovery of X-rays by Roentgen in 1895, they were being used to aid surgeons in carrying out operations.

Although small exposures to X-rays are harmless larger doses are harmful and ultimately lethal. Some X-rays not only kill cells but they can promote the development of cancer and some early researchers died in this way.

X-rays are produced in high voltage discharge tubes but certain radioactive substances also emit waves of the same wavelength. We call the rays from radioactive sources gamma rays (γ rays) although they are basically X-rays.

Questions

1. Under what circumstances is the limited range of radio waves (a) an advantage, (b) a disadvantage?

2. Sea waves are mostly transverse—what propels a surfer?

3. If the sensation produced in your brain by the colour red is the same as that produced in someone else's brain by the colour green, could you detect this?

4. When a transverse wave travels along a rope, each particle moves only at right angles to the rope. What is it that moves along the rope?

5. In a monophonic record player, does it matter which wire goes on which terminal of the speaker? Does it matter on a stereophonic one?

Multiple Choice

1. The frequency of a S.H.M. depends on which two factors?
(a) amplitude, (b) restoring force, (c) displacement, (d) inertia.

2. What two quantities reduce when oscillation is damped?
(a) amplitude, (b) energy of oscillation, (c) inertia, (d) mass.

3. Resonance occurs when the applied frequency
(a) equals the natural frequency, (b) is undamped, (c) reduces to zero, (d) is critically damped.

4. Which of the following are longitudinal waves?
(a) light, (b) sound, (c) waves in strings, (d) X-rays.

5. Which has the longest wavelength?
(a) X-rays, (b) radio waves, (c) light, (d) infra-red waves.

6. Which waves can *not* travel through a vacuum?
(a) sound, (b) light, (c) ultra-violet rays, (d) X-rays.

7. Which electromagnetic waves can produce sunburn?
(a) red light, (b) infra-red radiation, (c) microwave, (d) ultra-violet.

8. What would be the typical range of ships' radar?
(a) 3 km, (b) 30 km, (c) 300 km, (d) 3000 km.

9. Which equations express the relation between period T and frequency f of a wave?

(a) $T = f$, (b) $T = f^2$, (c) $T = \dfrac{1}{f}$, (d) $f = \dfrac{1}{T}$.

10. Refraction is caused by a change in wave
(a) speed, (b) amplitude, (c) energy, (d) damping.

11. Place the colours in order of increasing wavelength
(a) yellow, (b) blue, (c) red, (d) green.

12. Which do you associate with higher frequency sounds from a guitar?
(a) thick strings, (b) thin strings, (c) high tension, (d) low tension.

Problems

1. State the relation between speed, frequency and wavelength of a wave. Calculate the speed of a sound of frequency 256 Hz and wavelength 1.3 m. [333 m/s]

2. The frequency of light of wavelength 4.20×10^{-7} m, is 7.14×10^{14} Hz. Calculate its velocity. [$c = 3.00 \times 10^8$ m/s]

3. Calculate the frequency of a vibration which repeats itself every $\frac{1}{5}$ of a second. [5 Hz]

4. Calculate the frequency of sound whose period is 0.0025 s. [400 Hz]

5. Calculate the wavelength of Radio 1 (VHF) given that it broadcasts at a frequency of 89.1 MHz. (Speed of radio waves is 3.0×10^8 m/s.) [3.37 m]

6. Calculate the frequency of Radio 4 (AM) given that it broadcasts on a wavelength of 1500 m. (Speed of radio waves is 3.0×10^8 m/s.) [200 kHz]

7. A vibrator oscillates at a rate of 150 vibrations a

minute and produces a transverse wave of wavelength 30 mm. Calculate the speed of the wave. [75 mm/s]

8. Some ripples on the surface of water are travelling at 50 mm/s. They cause a fisherman's float to move up and down three times every 4 s. Calculate the wavelength of the waves. [67 mm]

9. The diaphragm of a loudspeaker vibrating at a frequency of 1.0 kHz produces a sound wave in air of wavelength 340 mm. Calculate the speed of sound in air. [340 m/s]

10. An echo sounder is used at sea to receive sound reflected at the sea-bed. If a pulse of sound from the transmitter on the ship is received back after an interval of 0.01 s, calculate the depth of the sea below the ship. (Speed of sound waves in sea water is 1600 m/s.) [8 m]

11. An observer counted 8 seconds between seeing a flash of lightning and hearing the sound of thunder. If the speed of sound is taken as 350 m/s, calculate how far away the lightning flash originated. [2.8 km]

12. Calculate the speed of transverse waves in a string under a tension of 30 N if its mass per unit length is 1.2 $\times 10^{-3}$ kg/m. [158 m/s]

13. State the equation relating the speed of transverse waves in a stretched string. How would the speed of waves in a string be changed if the tension was increased four fold? [speed is doubled]

14. If transverse waves travel along a rope of mass 0.15 kg at 8.0 m/s and the rope is 3.0 m long, what is the tension in the rope? [3.2 N]

Multiple Choice Answers

1(bd), 2(ab), 3(a), 4(b), 5(b), 6(a), 7(d), 8(b) 9(cd), 10(a), 11(bdac), 12(bc).

16.
Sound

Sound is essential to mankind. It is the basis of speech, by which we communicate ideas and information to one another. It provides us with a ready awareness of events which we could not detect so quickly with any of the other senses. In the form of music, it contributes to our enjoyment of life.

Sound Perception

The sound of a grasshopper can be heard at a range of many metres, at which distance the displacement the sound causes in the air is no bigger than the diameter of a molecule. At the other end of the scale, the ear can tolerate a thunderclap, which is about 10^{10} times more energetic than the sound of the grasshopper. In addition to this wide range of energies, the ear is astonishingly sensitive to the complexity of sound waves and it can detect, for example, the different instruments of an orchestra playing simultaneously. Hearing is the last of our senses to become dormant when we sleep and the first one to function on waking. This very highly developed sense of hearing has played an important part in animal survival by enabling animals to catch their prey and avoid their enemies.

The Ear

The ear has three sections—the outer, the middle and the inner ear. (Fig. 16.1.) The outer ear funnels the sound waves to a membrane called the ear drum, which is made to vibrate. In the middle ear, the vibrations of the ear drum are transmitted to a smaller membrane at the entrance to the inner

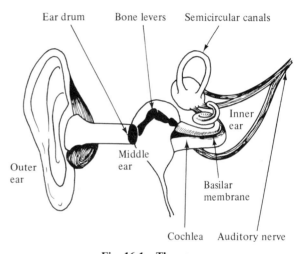

Fig. 16.1   The ear

ear. At the same time, the pressure variation is amplified about 60 times. The inner ear converts mechanical vibrations to nerve impulses.

The ear is situated near the semicircular canals which, by being sensitive to fluid pressure, give animals a sense of balance. When they are subject to unfamiliar signals for example on a boat or a fairground roundabout they can produce sea sickness or dizziness.

Sound Waves

The source of sound is vibration. The essential requirement for the propagation of sound waves is a medium in which the particles are close enough to react on each other.

This is provided by solids, liquids and gases, but not by a vacuum. The fluctuating quantity in a sound wave is the position of the particles in its

169

path, which cause periodic variations in the pressure. (Fig. 16.2.) Some sound vibrations you can see or feel, as is the case with a drum or a plucked guitar string. By plotting a graph of the displacement against the distance along a sound wave we can represent the differences between one sound and another. (Fig. 16.3.) Remember that sound waves are longitudinal waves, i.e., vibrations to and fro in the direction of wave travel. A sound wave is described by three major characteristics: loudness, pitch and quality. Each of which your ear can detect easily.

Loudness   This describes the amount of energy in the sound. The greater the movement of the particles, the more energy involved and the louder the sound. (Fig. 16.4.)

Pitch   The frequency of vibration determines the pitch. A high pitched sound such as a scream has a high frequency, i.e., many vibrations per second. (Fig. 16.5.)

Quality   Another name for quality is tone, and it describes the complexity of the sound. A piano and a trumpet playing notes of the same loudness and pitch sound differently. The difference is best illustrated by the graphs of two sounds. (Fig. 16.6.)

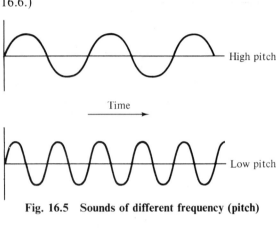

Fig. 16.5   Sounds of different frequency (pitch)

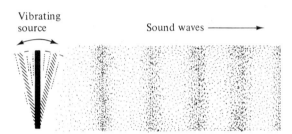

Fig. 16.2   Pressure variations in a sound wave

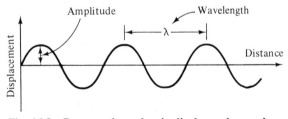

Fig. 16.3   Representing a longitudinal sound wave by a graph

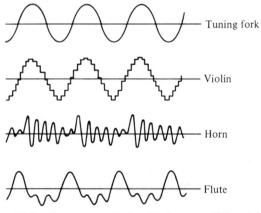

Fig. 16.6   The ear can distinguish between different instruments playing notes of the same frequency and loudness

Loudness Levels

For a given frequency, the more intense the note, the greater the stimulus to the cochlea and the louder the note is. Loudness however is a subjective measure and the perception of it varies from person to person.

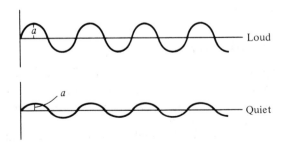

Fig. 16.4   Loud sounds have a greater amplitude (a) than quiet sounds

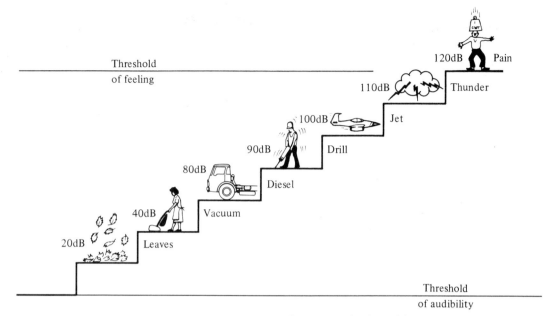

Fig. 16.7 Some examples of sound relative intensities

When a sound is doubled in energy an observer will by no means judge it to have doubled in loudness. In fact, about a tenfold increase in energy is needed to double the loudness. The response of the average ear to sound energy is logarithmic rather than direct proportion. For this reason we use a logarithmic scale of measurement for loudness. All sounds are compared with the quietest sound we can hear $[10^{-6} \, \mu W/m^2]$. This threshold of audibility is the zero of the decibel scale. (Fig. 16.7.) One decibel is about the smallest difference of loudness you can judge. Ten decibels is about a doubling of loudness and 120 decibels is the range from the quietest sound you can hear to the loudest sound you can tolerate.

Hearing is damaged by loud sounds. A very lound bang, e.g., an explosion can produce deafness instantly. Exposure to sound levels above about 85 decibels over a long period can also damage hearing. Weavers are often deafened prematurely by the noise of the shuttles. It is feared that the loudness of some disco music may have a similar effect on unsuspecting audiences.

Frequency Range

Below 20 vibrations per second the ear is aware of separate sounds rather than of a note. Above about 20 000 Hz the sound becomes inaudible to adults. Children may hear sounds of higher frequency and dogs certainly can. With advancing years human sensitivity to upper frequencies falls. By middle age the upper limit might be 10 000 Hz and in old age the limit may have dropped so low that consonants cannot be heard. (Fig. 16.8.) Sound of frequency greater than the audible limit is said to be ultra-sonic and vibrations in this range are finding increasing use. (Page 75.)

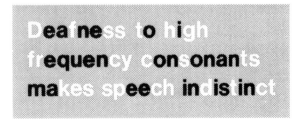

Fig. 16.8 An impression of the effect of partial deafness. Deafness to high frequency consonants makes only the black letter sounds audible

Reflection and Refraction of Sound Waves

When a sound wave meets a boundary at which its velocity changes, some of its energy is reflected. (Fig. 16.9.) The reflection is rarely regular, as for a

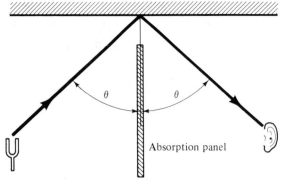

Fig. 16.9 **Sound is reflected regularly at a smooth hard surface; the maximum intensity being detected in the direction which makes angles of incidence and reflection equal**

light wave, because sound is strongly scattered by obstacles in its path.

When sound waves are changed in velocity by entering a different medium or by changes in the properties within one medium, their direction may be changed. This effect called refraction is occurring constantly to sound waves travelling through the air due to temperature gradients in the air. Thus, sounds carry further at night than during the day because of the different temperature distributions which exist at these times. (Fig. 16.10.)

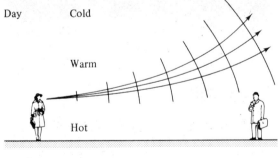

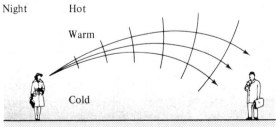

Fig. 16.10 **The typical daytime temperature gradient refracts the sound waves up into the air over the head of the listener. At night, the sounds are refracted down onto the ground and they are heard at a greater distance**

Doppler Effect

Everyone has heard the Doppler effect at some time, though perhaps without realizing it. The Doppler effect is the drop in the observed frequency of a source of sound as it passes and recedes from the observer. It is most evident in police or ambulance sirens as they pass at speed.

The effect is caused by a shortening of the wavelength of the sound in front of the moving source and a lengthening of the wavelength behind it. (Fig. 16.11.)

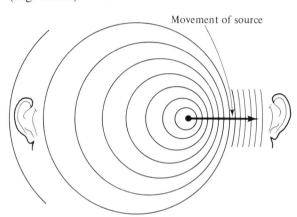

Fig. 16.11 **The shortening of wavelength before, and lengthening behind, a moving source changes the observed frequency**

The Doppler effect is also apparent to a moving observer who is passing a stationary source. This effect is due to the change of the relative velocity of the sound waves as they appear to the observer, the observed frequency being proportional to the relative velocity.

The Doppler effect also applies to light waves, producing, for example, a change in the colour of the light emitted from receding stars. The change in wavelength is towards longer wavelengths, that is, the red end of the spectrum, and this red 'shift' is used to estimate the velocity of stars. The effect is also used with radar waves to estimate the velocity of aircraft or the speed of cars.

Standing Waves

A wave which is reflected at right angles to a boundary travels back along its own path and produces an interference effect, which we call a standing wave. All points on a standing wave do not have the same amplitude as in a progressive wave. At certain fixed points (nodes) there is no movement hence the name standing wave. The amplitude increases from zero at the nodes to a maximum at the antinodes halfway between the nodes. So we have:

1. Nodes points of zero displacement
2. Antinodes points of maximum displacement

Standing waves are formed when waves confined to one direction are efficiently reflected as in a pipe or a rod or string. The formation of a standing wave is closely connected with resonance or vibrating objects. An object resonates at those frequencies for which the standing waves fit exactly into the dimensions of the object.

Resonance in Strings

Strings vibrating transversely often act as a source of sound. The factors which determine the frequency are summarized by the laws of vibration of a stretched string.

1. $f \propto \dfrac{1}{\text{length}}$

2. $f \propto \sqrt{\text{tension}}$

3. $f \propto \dfrac{1}{\sqrt{\text{mass of string}}}$

The ends of the string are fixed and must be nodes.

The simplest mode of vibration, and the one which accounts for the greatest part of the energy in a string, has a single antinode. In this case, the wave vibrates at its fundamental frequency f_1. (Fig. 16.12(a).)

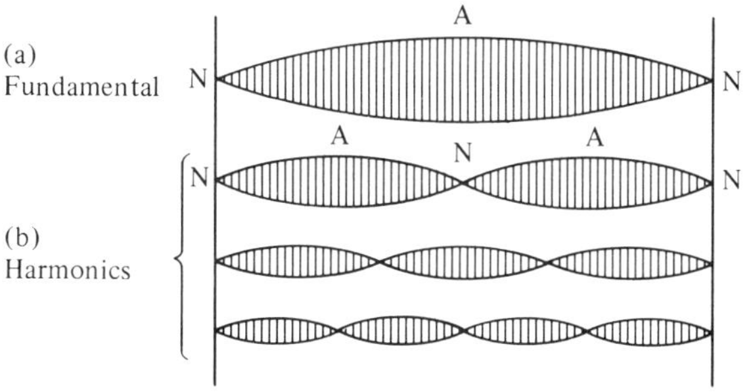

Other patterns or modes of vibration which produce nodes at the supports are possible. (Fig. 16.12(b).) These frequencies are present when the string is vibrating, but their amplitudes are much smaller than the fundamental frequency. (Fig. 16.13.) These secondary frequencies are called harmonics, and they are classified according to their ratio to the fundamental frequency. Thus frequencies of $2f_1$, $3f_1$, and $4f_1$ are the second, third and fourth harmonics respectively of the frequency f_1, which is itself the first harmonic. Plucking the string in the middle tends to favour the odd harmonics, which have an antinode in the middle of the string. A higher proportion of the energy is contained in the higher harmonics when the string is plucked nearer to the support. The difference in the sound of the string plucked in these two positions can be clearly heard.

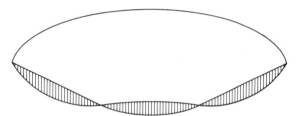

Fig. 16.13 Several waves may be present in a string at the same time

Stationary Waves in Air Columns

A column of air may be made to vibrate by means of a jet of air blowing against a sharp edge at one end of the pipe. (Fig. 16.14.) The disturbance acts as an antinode, sending waves up the column, where they are reflected, producing a stationary wave with a node at the rigid end. Only those waves that fit into the tube with an antinode at one end and a node at the other resonate and produce audible sound. The fundamental frequency and other frequencies emitted by the pipe are shown in Fig. 16.15.

When the end of the pipe is open, sound waves are still reflected. A compression reaching the open end is suddenly released producing a large particle amplitude or antinode at the end. The frequencies which resonate in an open-ended tube are shown in Fig. 16.16.

(a) Fundamental

(b) Harmonics

Fig. 16.12   Modes of vibration of a string under tension

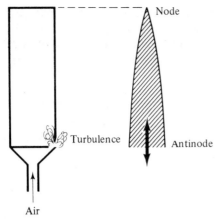

Fig. 16.14 The air in a pipe is set into vibration by the turbulence produced as a jet of air meets a sharp edge. The shaded diagram is a way of representing the vibration shown by the arrows at the antinode

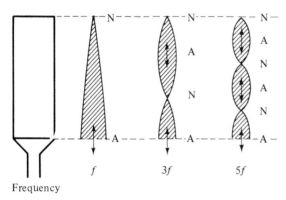

Fig. 16.15 Possible vibrations of the air in a closed pipe with a node at one end and an antinode at the other

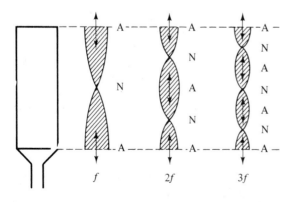

Fig. 16.16 Possible vibrations of the air in an open pipe with an antinode at each end

Musical Instruments

Wind instruments, such as organs and whistles, depend on the resonance of air columns. The pitch of a note emitted by a column is generally controlled by changing the length of the resonating column. In woodwind instruments, the physical length is fixed, but their effective length is varied by opening and closing holes and valves. The organ has a large number of pipes, which can be separately selected. In brass instruments, we increase their length by opening valves which introduce new sections of tube, and the trombone has a telescopic u-shaped tube which can slide in and out.

The resonance in some instruments, such as clarinets and saxophones, is controlled by reeds, while in others such as the trumpet, the vibration of the player's lips controls the resonance.

Acoustics of Buildings

The study and design of building interiors to produce specific listening conditions is called acoustics.

The acoustics of a building are closely linked with *reverberation*, that is, the persistence of a sound due to its repeated reflection from the boundaries of an interior. The time interval between the making of a sound and the moment when it just becomes inaudible is the *reverberation time*. It is usually defined as the time in which a note of 512 Hz falls to 1 millionth part of its initial intensity. The reverberation time is almost zero in the open air where there is no reflection. It is several seconds in a large swimming pool and, thus, sounds are indistinct because of the overlapping of reverberations. The optimum reverberation time is greater for orchestral music than it is for light music and speech. The ideal reverberation time thus might range from half a second for speech to give clarity to $2\frac{1}{2}$ seconds for orchestral music in a concert hall to give a resounding quality.

Geophysical Prospecting

An earthquake or major explosion generates vibrational waves which may travel very large distances through the earth. When these vibrations are recorded on seismographs at different places on the earth's surface, they can be used to detect and locate and classify the disturbances or to give information about the structure of the earth. The reflection of such waves through rock strata is used by geologists, prospecting for oil, to detect the structures likely to contain oil deposits.

Depth sounding of the ocean employs a similar technique. An instrument records the time between the sound and the return of its echo from the ocean bed. From a knowledge of this time and the velocity of sound in sea water, the depth can be deduced.

Ultrasonics

When a piece of quartz cut along a certain axis is compressed, positive and negative charges are exhibited on its opposite faces. The polarity of the charges reverses when the crystal is under tension. This is the *piezo-electric* effect. The alternate compressions and stretchings of a quartz crystal, when under the influence of a compressional wave, generate a small alternating potential across the crystal, which provides an effective means of detecting the wave.

Conversely, if we apply an alternating electrical potential to the quartz, it expands and contracts at the same frequency and acts as a source of waves. The quartz crystal is an efficient transducer, i.e., a device for converting electrical impulses to mechanical vibrations and vice versa.

Ultrasonic waves find many applications. Square or other shaped holes can be bored into a surface if a hardened rod, in contact with the surface, is subject to ultrasonic vibrations. The cutting head does not rotate, it vibrates and, with the help of an abrasive, produces holes the same shape as the cutting head. Plastic components can be welded together by ultrasonic vibrations which produce heat through friction at the points of contact.

Ultrasonic waves in liquids produce cavitation. That is, they produce tiny spaces in the liquid. The partial vacuum in these spaces exerts a strong pull on exposed solid surfaces, detaching any particles of dust which may be adhering to them. Even the insides of hypodermic needles can be cleaned in this way.

Ultrasonic vibrations in liquids may be used to emulsify immiscible liquids or to destroy bacteria by bursting their cells. These vibrations also have the effect (rather like an increase in temperature) of decomposing some liquids and of speeding up chemical reactions in others. The waves can be used to remove the air bubbles from liquids such as the molten metal used for castings.

One of the earliest ones was to improve depth sounding methods used by ships. The waves can be used to transmit messages between submarines, and they serve in a similar way to radar in tracking underwater craft and objects such as icebergs.

Another application of the echo sounding principle is the use of ultrasonic waves to detect flaws in large metal castings. Even invisible cracks in the metal can be detected by the waves they reflect back to the detector.

The thickness of a metal plate at any point can be measured even when we have access to only one side. The method allows us to check the thickness of plates, pipes, or sheathing, which are subject to corrosion, and it has even been used to measure the thickness of the fat layer on pigs.

Sound Reproduction

The past 100 years have seen a rapid growth in the importance in our daily lives of sound reproduction. Listening to sound recordings and broadcasts now occupies a major place in our leisure activities, and large sections of the entertainment and electronic industries are devoted to it.

The reproduction of sound relies on the conversion of sound vibrations into exactly corresponding vibrations of another quantity. The most convenient quantity into which sound can be converted is that of oscillating electrical current. We will describe separately three stages of the reproduction of sound, its conversion to an alternating current, the methods of regenerating these currents into sound and the methods of recording and storing the sounds.

Microphones

Microphones convert mechanical sound waves into electrical impulses. The qualities we require in a microphone are:

(a) A large electrical output.
(b) The faithful reproduction of audible frequencies.
(c) An absence of background noise.

We also want the microphone to be cheap and robust. (Fig. 16.17.) Shows one of several types.

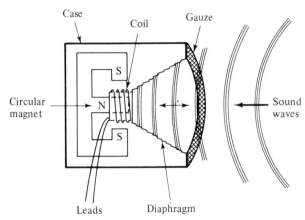

Fig. 16.17 Moving-coil microphone. The sound vibrating the coil generates an alternating electric current

Regenerating the Sound

Electrical impulses leaving the microphone are suitably amplified before being fed to a device for converting them back into sound vibrations. This is achieved by means of a speaker for powerful sound projection or an earphone for individual listening.

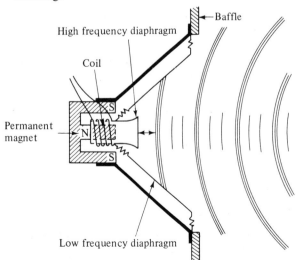

Fig. 16.18 Moving-coil loudspeaker

The moving-coil loudspeaker uses the principle of the moving-coil microphone in reverse. The alternating current is fed to a coil situated in the field of a magnet and attached to a diaphragm. (Fig. 16.18.) This produces a vibration in the cone-shaped diaphragm which, being large, can generate very intense sound waves.

Sound Recording

The gramophone disc is the most common means of sound recording for reproduction in our houses. At the recording studio, the output of a microphone, after being amplified, acutates an electromagnetic vibrator attached to a cutting tool. (Fig. 16.19.) The tool is in contact with a rotating wax disc, into which the tool cuts a wavy line as it vibrates. As the disc turns, the head traverses from its outside edge towards the centre, producing a long spiral groove. A metal impression is made from the wax disc which, in turn, is used to stamp out plastic replicas of the original disc. A record-player reverses the recording process. (Fig. 16.20.)

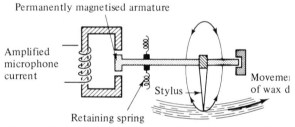

Fig. 16.19 The electromagnetic disc cutting head. The current gives a rotational vibration to a cutting stylus

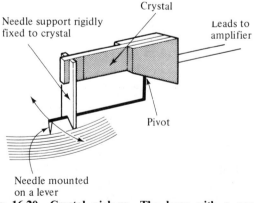

Fig. 16.20 Crystal pick-up. The lever with a needle attached is a push-fit in the pivot and it can quickly be replaced

Sound vibrations can be recorded as a pattern of varying degrees of magnetization on a plastic ribbon impregnated with a magnetic material. The tape moves at a constant speed across the poles of an electromagnet, which are less than 0.1 mm apart. The electrical output from a microphone is supplied to the electromagnet and controls the magnetic field between the poles. Thus, the section of tape bridging the poles at any instant is magnetized to a degree controlled by the sound vibration. (Fig. 16.21.)

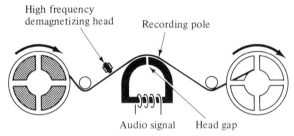

High frequency demagnetizing head

Recording pole

Audio signal   Head gap

Fig. 16.21   Tape recorder

During playback, the magnetized tape recording is moved at the same speed past the poles of the magnet and induces a field in the magnet. This, in turn, induces a current in the windings, which can be amplified to reproduce the original sounds.

Questions

1. What is the difference between progressive and standing waves?

2. How would you construct a lens to focus sound waves?

3. What is the difference between a musical sound and a noise?

4. Why is it important that the motors used in record players and tape recorders should run at very constant speed?

5. What qualities are required for a good pick-up and a good loudspeaker?

Multiple Choice

1. Which quantity indicates the pitch of a note?
   (a) loudness, (b) frequency, (c) tone, (d) resonance.

2. Which quantity would be increased by an increase in decibels?
   (a) frequency, (b) loudness, (c) pitch, (d) quality.

3. What change in the sound from an advancing source is the most apparent as it passes the receiver?
   (a) loudness, (b) pitch, (c) speed, (d) intensity.

4. If two instruments sound notes of the same pitch in what ways could they differ?
   (a) frequency, (b) velocity, (c) wavelength, (d) loudness.

5. Which correctly relates frequency f to wavelength λ?

   (a) $f \propto \lambda$, (b) $f \propto \dfrac{1}{\lambda}$, (c) $f \propto \lambda^2$, (d) $f \propto \dfrac{1}{\lambda^2}$.

6. Which does not affect the frequency of vibration of a string?
   (a) length, (b) tension, (c) amplitude, (d) mass.

7. Which of the quantities is zero at a node on a vibrating string?
   (a) wavelength, (b) displacement, (c) tension, (d) amplitude.

8. Compared to sound waves ultrasonic waves have
   (a) higher frequency, (b) longer wavelength, (c) greater audibility, (d) lower frequency.

9. The reverberation time is the time taken for a sound intensity to reduce to

   (a) $\frac{1}{2}$, (b) $\frac{1}{10}$, (c) $\frac{1}{100}$, (d) $\frac{1}{1\,000\,000}$.

10. The light received from a receding star is different from an identical stationary star in that it is
    (a) more red, (b) more blue, (c) brighter, (d) dimmer.

Problems

1. A sonar beam is radiated from a ship and an echo is received after 0.9 s. If the speed of sound in the water is 1400 m/s, how far away is the reflecting surface? [630 m]

2. How does the frequency of vibration of a stretched wire depend on (a) the tension, (b) the mass per unit length, (c) the length. A wire vibrating at 250 Hz has its tension doubled. What is the resulting frequency? [354 Hz]

3. What length would an open organ pipe have to be to give a frequency of 500 Hz? (Speed of sound = 330 m/s.) [0.33 m]

4. If a stretched string 0.6 m long is plucked at its centre, what is the wavelength and frequency of the fundamental note? (Speed of sound in string = 840 m/s.) [1.2 m; 700 Hz]

5. A steel wire 0.5 m long and mass 4.0 g is stretched with a tension of 500 newtons. What is the frequency of the fundamental note emitted by the wire? [250 Hz]

6. A steel wire of length 239 mm and diameter 0.80 mm vibrates with a fundamental frequency of 256 Hz. What is the tension in the wire? (Density of steel = 7800 kg/m^3.) [58.7 N]

7. Someone on a station platform notices that the apparent frequency of a train's whistle drops from 900 to 800 Hz as the train passes through. How fast is the train going? (Speed of sound = 330 m/s.) [19.4 m/s]

8. A stationary source emits a note of 500 Hz. What frequency does an observer hear if he approaches it at 30 m/s and then goes away from it at the same speed? (Speed of sound = 330 m/s.) [545 Hz; 455 Hz]

9. What is the fundamental frequency produced by blowing across the edge of a closed tube 10 mm long? (Speed of sound = 330 m/s.) [8250 Hz]

10. What are the wavelengths in air of the piano's lowest note (frequency 27.5 Hz) and its highest (frequency 4186 Hz)? (Speed of sound = 330 m/s.) [12 m; 79 mm]

Multiple Choice Answers

1(b), 2(b), 3(b), 4(d), 5(b), 6(c), 7(bd), 8(a), 9(d), 10(a).

17.
Optics

Geometrical optics is the study of the production of images and is based on the assumption that light travels in straight lines. We can see that the assumption is approximately true because point light sources cast sharp shadows and spotlights throw well defined beams.

Reflection at Plane Surfaces

Light meeting a surface is only partly reflected, some is transmitted and some is absorbed. You will understand this better by thinking of cases where one or other process predominates, for example, a blackboard, a window, or a white painted wall or a mirror.

In considering the reflected component, we might pause to question the difference between the reflection at a matt white and at a mirror surface. The difference is not in the amount of light they reflect, it is in the way the light is scattered on reflection as shown in Fig. 17.1. The matt surface

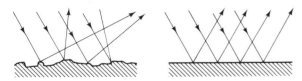

Fig. 17.1 Diffuse and specular reflection

reflects diffusely while the mirror reflects regularly. It is this latter type, called specular reflection, which we shall be concerned with. The direction of a ray is always defined by the angle it makes with the normal to the surface and not with the surface itself. (Fig. 17.2.) The symmetry apparent in the reflection of a ray at a plane surface is summarized in the two laws of reflection:

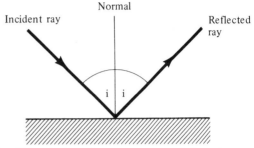

Fig. 17.2 Angles of rays are measured from the normal to the reflecting surface

1. The angle of reflection is equal to the angle of incidence.
2. The reflected ray is in the same plane as the incident ray and the normal.

The second law means simply that the reflected ray is not turned to left or right of the normal as viewed from the incident ray. An important application of the laws of reflection is made in the design of some electrical instruments. The scale of a galvanometer is easier to read if the scale is enlarged and the pointer is lengthened to match. A sensitive movement might be overloaded by a heavy pointer, but if the pointer is replaced by a minute mirror which moves with the mechanism, a ray reflected from it can be as long as required without increasing the inertia of the system. (Fig. 17.3.) As a bonus, the magnification is further increased by the fact that the total angle of deflection of the reflected beam is twice the angle of deflection of the suspended mirror.

Image Formation in a Plane Mirror

When rays from the object obeying the laws of

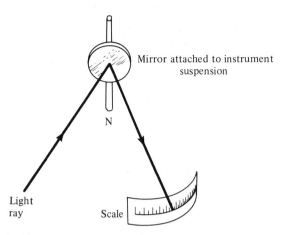

Fig. 17.3 **The optical lever. A very long light beam may be used as a pointer without adding weight to the suspension**

reflection are constructed, they diverge after reflection at the mirror. By covering the mirror in Fig. 17.4 you will see that the eye receives light as though from a point behind the mirror. The triangles PBO and PBI are congruent making PO = PI. Thus, the image is as far behind the mirror as the object is in front, a fact which is a matter of everyday observation.

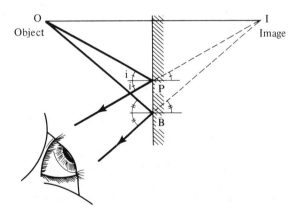

Fig. 17.4 **The image is as far behind a plane mirror as the object is in front. The image exists at I in the sense that the eye receives a cone of light diverging from I**

Note also that the image in a plane mirror is reversed. (Fig. 17.5.) If you are not convinced by the diagram, try to read a book reflected in a mirror.

An important aspect of any image in optics is whether or not it can be received directly on a

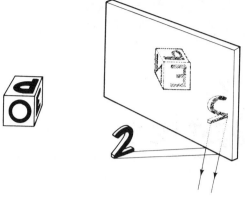

Fig. 17.5 **Lateral inversion of the image**

screen. The light rays from the plane mirror (Fig. 17.5) do not come from the image I, although they indicate its position. The image would not be visible on a screen placed at I and for this reason it is called a virtual image. The image viewed in a microscope is of this type.

Any image which can be seen on a screen placed in the position of the image is called a real image. Projected images such as those on a cinema or television screen are real images.

Reflection at Curved Surfaces

The giant radio telescope at Jodrell Bank and the small electric pocket torch, both employ curved reflectors. One is an enormous bowl, 750 t in weight, 250 feet in diameter, which gathers radio waves from outer space and concentrates them onto a focus so that they can be studied and analyzed. (Fig. 17.6.) The other is a small metal cup, only an inch or two across, which receives the light from a small bulb, and reflects it outwards to provide a convenient source of illumination. We will consider the three basic curved reflectors in common use: the concave spherical mirror, the convex spherical mirror, and the parabolic mirror.

A projectile fired upwards at an angle to the vertical describes a curve called a parabola. (Fig. 17.7.) The surface formed by rotating this curve about its axis of symmetry has the same shape as a parabolic reflector. The special feature of a parabolic reflector is that it concentrates a wide beam of light, travelling parallel to the principal axis of

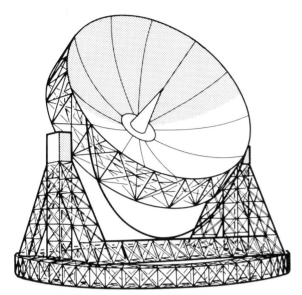

Fig. 17.6 The radio telescope at Jodrell Bank, which was built to detect radio waves from outer space, has a parabolic reflector 250 ft in diameter

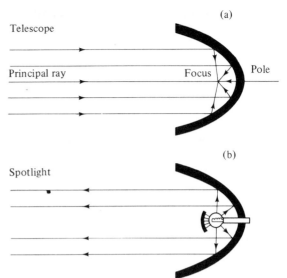

Fig. 17.8 Two uses of the parabolic reflector

Spherical Mirrors

A section of a spherical shell silvered on the inside can also produce a sharp focus from a parallel beam of light. (Fig. 17.9(a).) The distance between the mirror and the focal point is the focal length f.

the reflector, onto a single focal point. Such reflectors are used in some types of telescope. (Fig. 17.8(a).) But, more commonly, a source of light is placed at the focus and the reflector produces a parallel beam as in a car headlight. To prevent any divergence of the direct light from the bulb, a shield is placed in front of the bulb, as in a spotlight. (Fig. 17.8(b).)

A parabolic mirror can only produce an image if the light rays from the extremes of the object are almost parallel, as is the case with rays of light from distant objects like stars and satellites.

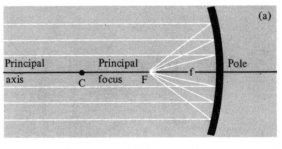

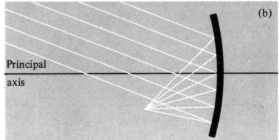

Fig. 17.9 A parallel beam of light in the direction of the principal axis passes after reflection through the principal focus. A parallel beam at an angle to the axis is brought to a focus off the axis

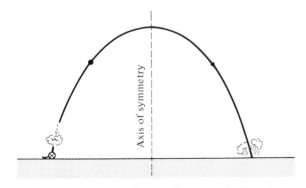

Fig. 17.7 A projectile describes a parabolic path

The focal point occurs at half the distance from the mirror to the centre of curvature, i.e.,

$$\text{Radius} = 2f$$

If the beam of light is at an angle to the principal axis it still produces a focus. (Fig. 17.9(b).) This is an important characteristic, because it makes possible the formation of an image of an object which subtends an appreciable angle at the mirror.

Spherical mirrors produce images just as plane ones do, but the position of the image is not quite so easy to predict. A diverging cone of light from a point on an object (situated outside the centre of curvature of a concave mirror) (Fig. 17.10(a)) is changed after reflection to a converging cone. The apex of this cone is a point on the image. Many such cones make up the complete image but its position can be located by tracing two particular rays from the top of the object. (Fig. 17.10(b).)

1. The ray parallel to the principal axis passing through the focus after reflection.
2. The ray directed through the centre of curvature meeting the mirror normally and returning along its own path.

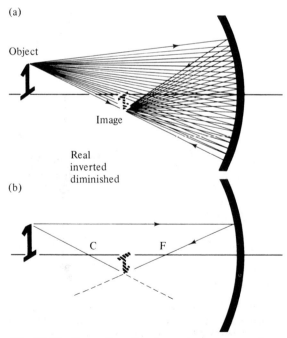

(a)

Object

Image

Real
inverted
diminished

(b)

C   F

Fig. 17.10   Formation of an image in a concave mirror

At the point of intersection of these two rays is the image of the top of the object. The complete image extends from this point towards the principal axis.

Nature of the Image

Three facts are necessary to describe an image: whether it is (a) real or virtual, (b) magnified or diminished, (c) upright or inverted. Thus, the image (Fig. 17.10) is real because it would be projected on a screen placed at I, it is diminished being smaller than the object, and it is, of course, inverted.

As the object is moved towards the mirror, the image position changes as shown in Fig. 17.11. The two construction rays can be used to locate the image, even when these particular rays would not in fact encounter the mirror.

The Convex Mirror

If a spherical shell is silvered on the outside, the result is a convex mirror. Light incident on the mirror parallel to the axis is reflected away from a focal point situated behind the mirror. (Fig. 17.12.) The focal length of the mirror is equal to half the radius of curvature. A convex mirror produces an image behind the mirror which is virtual, diminished, and upright for all positions of the object. (Fig. 17.13.) An object at infinity has its image at the focal point of the mirror. As the object moves from infinity towards the mirror, its image moves from the focal point towards the mirror. Convex mirrors are popular as driving mirrors, because they give a wide angle of view. (Fig. 17.14.) However, because of the small movement of the image for a large movement of the object, it is rather difficult to estimate distances in convex mirrors.

Refraction

When light passes from a vacuum into a material it slows down. This reduction in speed causes the light to bend at the surface. (Fig. 17.15.) Note how the light waves are slowed down at one end where they meet the surface causing the waves to change direction. The rays bend towards the normal when entering a denser medium and away from the normal when leaving.

This bending called refraction is responsible for such natural phenomena as mirages and rainbows. It enables lenses to magnify and thus provides the basis of optical instruments ranging from simple spectacles to the most complex wide screen projector.

Laws of Refraction

The behaviour of the ray at the surface can be described by two simple laws.

1. The ratio of the sine of the angle of incidence to the sine of the angle of refraction is constant.
2. The incident ray, the refracted ray and the normal at the point of incidence lie in one plane.

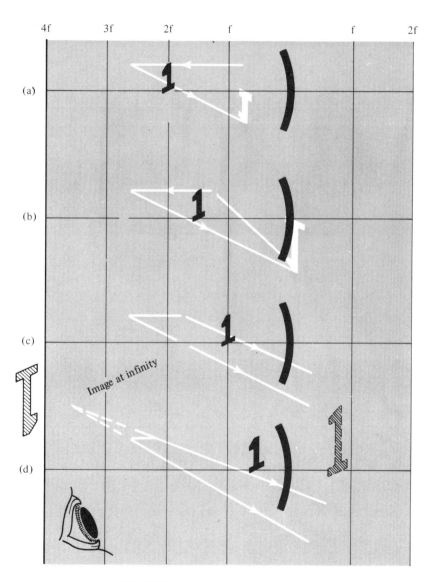

Fig. 17.11   Images in a concave mirror

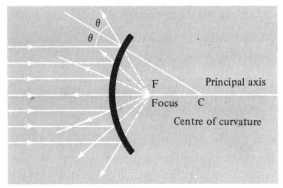

Fig.17.12 Effect of a convex mirror on a parallel light beam

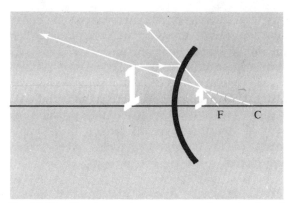

Fig. 17.13 The image in a convex mirror is virtual, diminished and upright

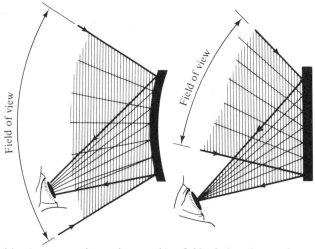

Fig. 17.14 A convex mirror gives a wider field of view than a plane mirror

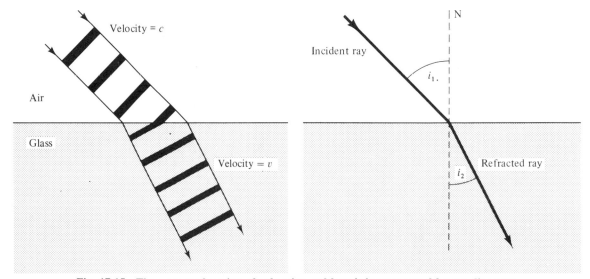

Fig. 17.15 The wave explanation of refraction and how it is represented by ray diagrams

The first law follows directly from the wave diagram. (Fig. 17.16.) During the time t, the end B of the wave front moves a distance ct while end A moves a distance vt.

$$AD \sin i_1 = ct$$
$$AD \sin i_2 = vt$$
$$\therefore \frac{\sin i_1}{\sin i_2} = \frac{c}{v} = \text{a constant}$$

The second law simply means that the refracted ray does not deviate to left or right of the plane of incidence.

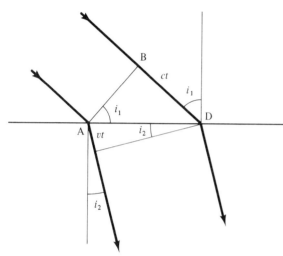

Fig. 17.16   Wave explanation of law of refraction

Refractive Index

The ratio of the velocity of light in a vacuum c, to that in a particular medium v, is called the refractive index of the medium n

$$n = \frac{c}{v}$$

Alternatively, from above equation

$$n = \frac{\sin i_1}{\sin i_2}$$

since c is the maximum velocity at which energy can travel in the universe, this ratio is greater than one.

Note that the refractive index correlates with the density of the material. (Table 17.1.) For this

Table 17.1   A Comparison of Refractive Index and Density ($\lambda = 5.893 \times 10^{-7}$ m)

Substance		Density (g/cm^3)	Refractive index
Water		1.0	1.33
Vitreous silica		2.2	1.46
Quartz		2.7	1.54
Glass—crown		2.5	1.48
	flint	3.0	1.56
	denser crown	3.6	1.61
	denser flint	4.76	1.74
Diamond*		3.52	2.42

* Diamond has the greatest refractive index of these substances but not the greatest density.

reason, refractive index is sometimes called optical density. Because air has a low density the velocity of light in air is approximately the same as that in a vacuum.

One effect of refraction is to make submerged objects appear nearer to the surface than they really are. Thus a partially submerged object appears bent at the surface and an object viewed through a slab of material appears to be displaced towards the slab. (Fig. 17.17.)

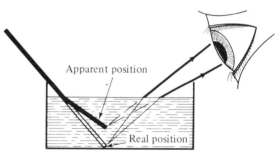

Apparent position

Real position

Fig. 17.17   A partly submerged object appears to be bent at the surface, due to refraction of the light rays

Total Internal Reflection

Consider a ray of light being reflected and refracted at a surface, as it enters a less optically dense material. (Fig. 17.18(a).) An increase in the angle of incidence causes the refracted ray to approach the surface until eventually it skims the surface. (Fig. 17.18(b).) A further increase in i and the refracted ray disappears. (Fig. 17.18(c).)

185

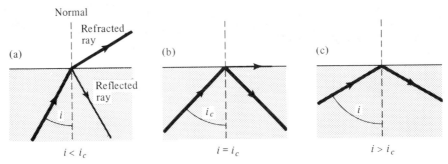

Fig. 17.18   The refracted ray disappears as the angle of incidence in the denser medium exceeds the critical angle i_c. The ray is then totally internally reflected

The laws of refraction still apply to the critical ray.

$$n = \frac{\sin 90}{\sin i_1}$$

$$n = \frac{1}{\sin i_c}$$

Problem

What is the maximum angle of incidence of light which emerges from the surface of glass of refractive index 1.5?

Solution

Light emerges for angles of incidence from zero to the critical angle i_c.

$$\sin i_c = \frac{1}{n} = \frac{1}{1.5}$$

$$i_c = 42°$$

Total internal reflection can be used to direct light into otherwise inaccessible places. Light directed into a rod of pliable material is confined to the rod by repeated total internal reflections and emerges only at the end of the rod where it meets the surface almost normally. (Fig. 17.19.) Bunches of very fine filaments can transmit a complete image back to the eye from an inaccessible place.

The heart can be viewed from the inside without surgery by passing a probe containing thousands of optical fibres through an artery in the arm. The valves and pistons of an engine can be examined without dismantling by a probe inserted through the spark plug hole. Optical fibres also convey signals in the form of fluctuating light beams over

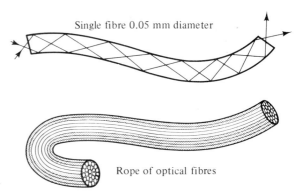

Fig. 17.19   A light probe directs light to or from inaccessible places by repeated internal reflections

distances more efficiently and more compactly than telegraph cables.

Prism Reflectors

Total internal reflection has two distinct advantages:

It is a very efficient form of reflection and
All the reflection occurs at a single surface.

Reflecting prisms are used in binoculars to extend the optical length while keeping the physical length as short as possible. Another example is the periscope, used in submarines to view the surface when submerged. (Fig. 17.20.)

Mirages

The gradual change in density of air due to a difference in temperature produces deviation of a ray passing through it. This can cause a mirage

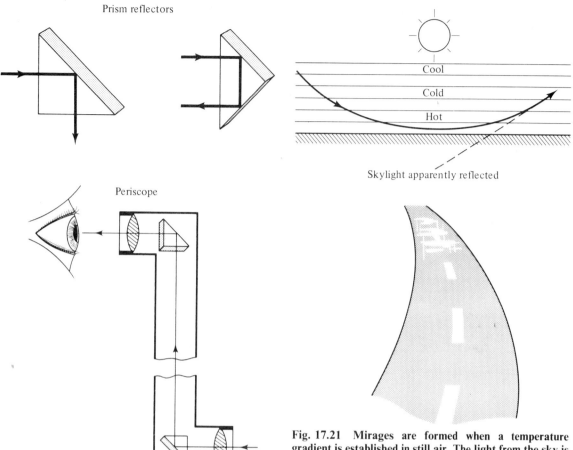

Prism reflectors

Periscope

Cool

Cold

Hot

Skylight apparently reflected

Fig. 17.20 45-degree angled prisms used as single or double reflectors are employed as reflectors in periscopes and some types of binoculars

Fig. 17.21 Mirages are formed when a temperature gradient is established in still air. The light from the sky is deflected upwards and appears to come from the surface of the ground

to be formed under certain conditions. These conditions are often realized on a road which has been heated by the sun and which establishes a vertical temperature gradient in the air above it. (Fig. 17.21.) The rays from the sky entering the layers of air of decreasing density are refracted away from the normal until eventually they are totally internally reflected upwards before reaching the road. The blue colour of the light from the sky gives the mirage the appearance of puddles on the road.

Measurement of Refractive Index

In forensic science, the science of crime detection, the measurement of refractive index can provide important evidence. Glass, being a mixture of substances, exhibits a variation of index from pane to pane. Minute fragments of glass found, perhaps, in a suspect's turnups or embedded in the fabric of a sleeve, may be identified with that of a broken window. The chance of two pieces of glass having the same refractive index to an accuracy of one part in 10 000 is remote unless they are from the same pane of glass.

Impurity in a substance may cause its refractive index to differ from that of the pure product. The difference in index is a sensitive measure of the amount of impurity present, and so the measurement of refractive index serves as a powerful tool for chemical analysis.

The most accurate methods of measuring the refractive index of a liquid are those which depend on the measurement of the critical angle. In the Abbé refractometer, the light is arranged to strike a liquid-glass interface at a range of angles, including grazing incidence. (Fig. 17.22.)

The light transmitted into the second prism makes a range of angles up to but not beyond the critical angle for the surface. The emergent rays from the prism are picked up by a microscope which is set up on the division between light and shade. The angular setting is dependent on the refractive index, and instruments are calibrated directly in refractive index.

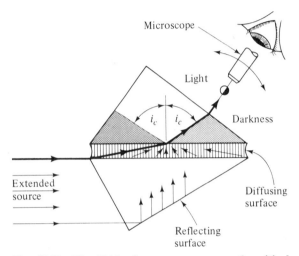

Fig. 17.22 The Abbé refractometer measures the critical angle of a liquid

Lenses

In most industrialized countries, the manufacture of optical instruments is booming. Movie cameras and slide projectors, once considered the equipment of the specialist, are becoming commonplace in our homes. The application of optical instruments to problems in engineering, from time and motion study to rapid accurate measurement, is increasing. The assessment of the value of an optical aid often boils down to an understanding of the optics of the lens, which is the basic element of most optical instruments.

A lens is usually bounded by two spherical surfaces and may have a variety of shapes. Basically, lenses fall into two categories, convex lenses which are thicker at the centre than at the perimeter, and concave lenses which are thicker at

the perimeter. Both convex and concave lenses may be bent into several shapes which although they look different have essentially the same power. (Fig. 17.23.)

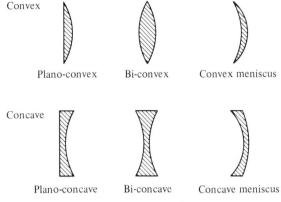

Fig. 17.23 Convex and concave lenses bent into different shapes

Effect of a Lens on Light Waves

You can appreciate the action of the two types of lenses by considering their effect on light waves. A series of plane light waves passing through a lens is slowed down in the lens material. In a convex lens, the centre of the waves spends more time in the lens and is therefore retarded, allowing the perimeter of the waves to overtake on the outside. (Fig. 17.24.) After leaving the lens, the waves converge to a point on the principal axis. This point is the principal focus of the lens.

A concave lens retards the waves most at the perimeter. The waves then travel outwards and diverge as though from a point on the incident side of the lens. (Fig. 17.24.) Because the waves do not originate at this point, we call it a virtual focus.

The focal length of a lens is the distance between the lens, and the focal point, F. A lens has the same focal length for either direction of incident light.

Formation of an Image by a Convex Lens

An illuminated object scatters light in all directions and so we can consider each point on the object to be a source of light.

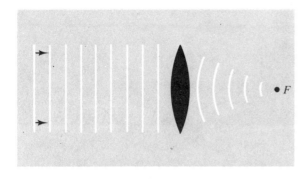

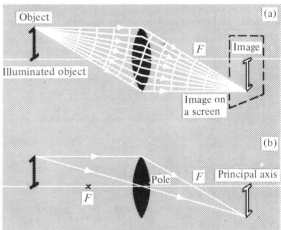

Fig. 17.25 The light waves can be more conveniently represented by rays. Two rays are required to locate an image. (a) A ray parallel to axis passing through *F*. (b) A ray passing through the pole of the lens

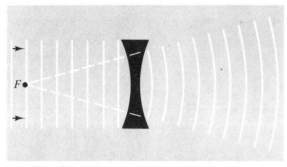

Fig. 17.24 A series of plane light waves is made to converge to a point by a convex lens and to diverge from a point by a concave lens

A cone of light from a point on the object is converged by the lens to a point, which forms part of the image. (Fig. 17.25.) The exact position of the image may be obtained by tracing only two rays from a point on the object. One ray parallel to the axis is refracted through the focus. A second ray in the direction of the pole (where the sides of the lens are parallel) passes through the lens undeviated. These two rays intersect at a point on the image. (Fig. 17.25.)

When the object is outside the focal length, the image is real and inverted. (Fig. 17.26(a).) When the object is within the focal length, the rays leaving the lens are still divergent. They appear to diverge from an upright virtual but magnified image on the same side of the lens as the object. (Fig. 17.26(e).)

When the object is at the focal distance, the emergent light is parallel. The image is said to be formed at infinity, it is magnified and regarded as real. (Fig. 17.26(d).) At an object distance of $2f$ the lens produces an image of the same size as the

object. (Fig. 17.26(b).) The diagram also shows that the minimum distance between a real object and image is $4f$, a fact well worth remembering in experimental work.

Formation of an Image by a Concave Lens

A ray from the object to the pole of a concave lens passes through undeviated. A ray parallel to the axis after refraction diverges from the focus. (Fig. 17.26(f).) The image formed at the intersection of these two rays is diminished, virtual, and upright. This description applies to all the images of a real object produced in a concave lens, although the magnification does vary with the object position. You can see the extent of this variation if you construct separate ray diagrams for a distant object and for an object almost in contact with the lens.

Optical Instruments

The purpose of optical instruments is to extend human vision.

The optical systems described here are basic ones which will help you to appreciate quickly the system of any instrument you meet.

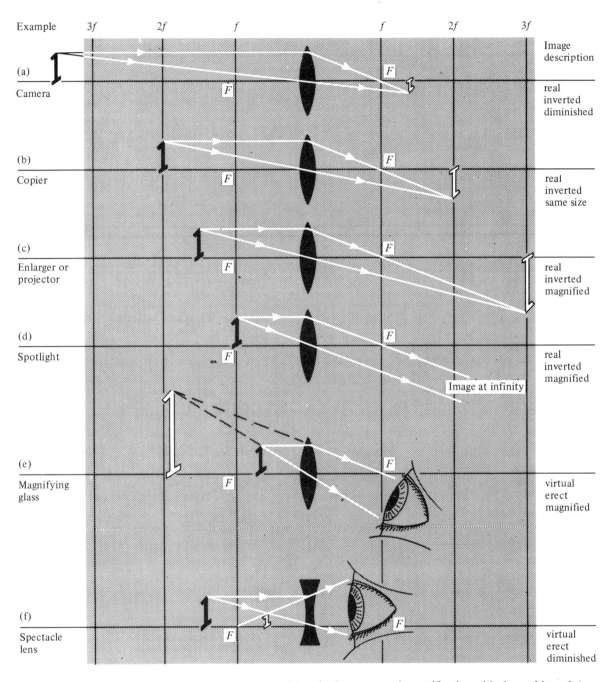

Fig. 17.26 The image produced by a convex lens (a)–(e) varies in nature and magnification with the position of the object. The image produced by a concave lens (f) is virtual, erect, and diminished for all positions of a real object

The Eye

The optical instrument with which we are most familiar is the eye. We usually think of the eye as the organ with which we see the world around us, but, in fact, the eye cannot see at all. It is the brain that actually sees, using the eye as its primary optical instrument.

Basically, the eye consists of a convex lens, which produces a real, inverted, and diminished image of the screen formed by the curved back of the eye-ball. (Fig. 17.27.) This surface, called the

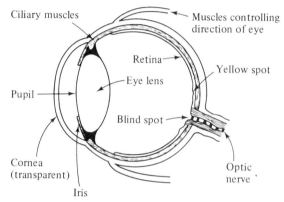

Fig. 17.27   This is a cross-sectional plan view of the right eye

retina, is covered with millions of nerve-endings, which are sensitive to light, and which transmit messages to the brain about the shape and colour of the image. The brain sorts out these messages, and from the small, upside-down image on the retina interprets the information it receives about the world outside. The eye, of course, like the camera, has to alter its focus according to the distance of the object it is looking at, the power of its lens being controlled by the *ciliary* muscles.

The eye is only sensitive to fine detail over a very small area, called the *yellow spot*, at the centre of the retina. This spot, which corresponds to a visual angle of only 1°, is rich in *cones*, nerve-endings which record the detail of an image.

The second type of light-sensitive nerves are the *rods*, which predominate in the rest of the retina. These rods are sensitive to a lower level of illumination than the cones, and we rely on them to see in poor lighting. We can, in fact, detect dimly-lit objects more clearly when we are not looking directly at them. The rods are also more sensitive

to movement than the cones, and this is why you can catch a moving ball an instant after you have seen it out of the corner of your eye.

The retina is protected from over-illumination by the iris, an opaque coloured membrane at the front of the eye. It responds to strong light by contracting the pupil, which is the aperture at the centre of the iris. Both the iris and the pupil are protected by a transparent layer of tissue, called the cornea, and the whole eye is kept clean by the eyelids which wipe it clean every few seconds.

The brain, of course, does not rely only on the information it receives from one eye, but considers the images formed simultaneously on the retinas of two eyes; and, because the two eyes are in different positions, the image received by one is slightly different from the image received by the other. By this means the degree of parallax occurring between objects at different distances is measured by the brain, which enables it to conceive the objects in three dimensions, and to estimate the distance of the objects. The binocular nature of our vision has been imitated in the stereoscopic camera, and has been given a more practical application in the range-finder.

The Magnifying Glass

An object placed within the focal length of a convex lens gives rise to an erect and magnified image further from the lens than the object. This enables the object to be viewed when it is much nearer the eye than would otherwise be possible. In this position the object subtends a greater angle at the eye and therefore produces a larger image on the retina. (Fig. 17.26(e).)

The Microscope

The microscope consists basically of two lenses of very short focal length. A minute object placed just outside the focal length of the objective lens produces a magnified real and inverted image. (Fig. 17.28.) This image falls inside the focal length of the eyepiece and hence produces a final image which is further magnified and remains inverted. Being virtual, this image could not be received on a screen at *I* and it is viewed by the eye which is placed near to the eyepiece.

191

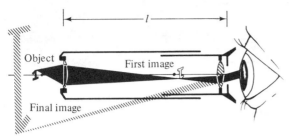

Fig. 17.28 The compound microscope produces a final image which is inverted virtual and magnified

Telescopes

The astronomical telescope had a basic optical system similar to the compound microscope. An important difference, however, is that the objective lens has a long focal length.

The final image of the telescope is laterally inverted and upside down. We can modify it for general use as shown in Fig. 17.29 when we call it a *terrestrial telescope.*

Binoculars are a modification of the astronomical telescope whereby an erect image is produced and the tube length of the instrument shortened. The light is totally internally reflected four times by two 45 degree prisms. (Fig. 17.30.) The larger rectangular surfaces of the prisms are parallel and their triangular surfaces are at right angles to each other.

Galileo, the first man to direct a telescope

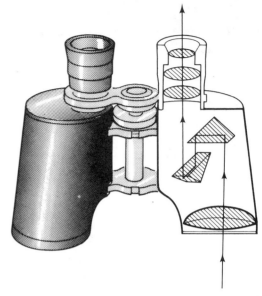

Fig. 17.30 The prisms in binoculars produce a final upright image and reduce the physical length of the tube

towards the stars, used a concave lens, as the eyepiece of his telescope. The system of the Galilean telescope is still used in opera glasses. (Fig. 17.31.)

Projectors

A projector has two distinct parts; the system of illumination and the projection lens. The large

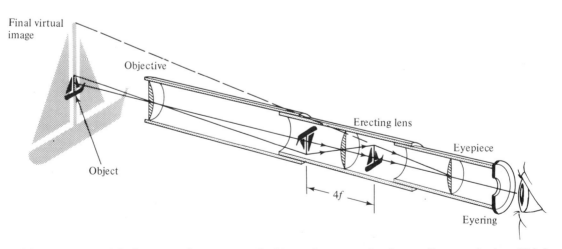

Fig. 17.29 The terrestrial telescope produces an erect final image by means of an intermediate erecting lens. This lens increases the length of the telescope by at least four times its focal length

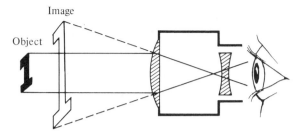

Fig. 17.31 Galilean telescope. The instrument produces an upright final image; its tube length is short but its field of view is small

magnification produced by projectors demand a very high level of illumination which is usually provided by a tungsten filament or carbon arc source. A large fraction of the total light from the source is directed onto the film by a curved mirror and condensing lens system. (Fig. 17.32.) This particular arrangement of the condenser lenses minimizes spherical and chromatic aberrations, giving an even white illumination of the film. This film lies just outside the focus of a lens which produces a real magnified image on the screen. The film itself is inverted and gives rise to an upright image.

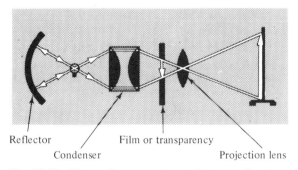

Fig. 17.32 The projector produces high magnifications and requires a very intensely illuminated object

The Camera

The simplest form of camera is a box with a pinhole at one end. (Fig. 17.33.) The image is made up of light from points on the object defined into narrow beams by the pinhole. If the pinhole is increased in area, the image becomes brighter but more diffuse. In order to obtain enough light to expose a photographic film adequately, a convex

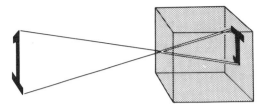

Fig. 17.33 The pinhole camera

lens is used. (Fig. 17.34.) The light energy falling on unit area of the film increases with the diameter of the lens but decreases with its focal length.

$$\text{Illumination} \propto \frac{(\text{diameter})^2}{(\text{focal length})^2} = \frac{d^2}{f^2}$$

The aperture controls of cameras are marked off at intervals giving a doubling of the illumination in each case.

$\dfrac{d^2}{f^2} \propto$ illumination $\propto$	$\dfrac{1}{500}$	$\dfrac{1}{250}$	$\dfrac{1}{125}$	$\dfrac{1}{63}$	$\dfrac{1}{32}$	$\dfrac{1}{16}$
f numbers (f/d)	22	16	11	8	5.6	4

The values of f/d, called f numbers, are marked alongside each of these divisions, called stops.

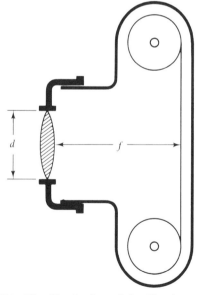

Fig. 17.34 The illumination of the film increases with the area of the lens and reduces with the lens to screen separation

Moving objects must be snapped in a short interval of time to prevent a blurred effect due to the movement of the image. This requires an increased size of lens aperture to provide adequate illumination of the film.

The lens can be moved to focus objects from about a metre distance to infinity. In general, we use as small an aperture as possible in order to increase the depth of field of the lens; i.e., the range of distances within which objects are in sharp focus.

Questions

1. Why is it unusual for a convex mirror to be used to project an image on a screen?

2. How does your image in a plane mirror differ from how other people see you?

3. Do you prefer a plane, concave or convex driving mirror? Why?

4. Suggest reasons why very powerful binoculars are not used for watching horse-races?

5. Why would you be suspicious if you measured a refractive index and found it to be less than 1?

Multiple Choice

1. Which of the following mirrors would give the greatest angle of view when used as a driving mirror?
(a) parabolic, (b) plane, (c) convex, (d) concave.

2. Which of the mirrors in question 1 produce real images?

3. What is the relation between the focal length f and the radius of curvature r of a concave mirror?

(a) $r = 2f$, (b) $r = f$, (c) $r = \dfrac{f}{2}$, (d) $r = \dfrac{2f}{3}$.

4. Which words describe the image in a plane mirror?
(a) real, (b) virtual, (c) laterally inverted,
(d) magnified.

5. Which words describe the image in a convex mirror?
(a) magnified, (b) virtual, (c) real, (d) diminished.

6. Which lenses would act as a magnifying glass?
(a) plano-convex, (b) plano-concave, (c) concave,
(d) convex meniscus.

7. A beam of light on entering a denser material is generally
(a) retarded, (b) deviated, (c) accelerated,
(d) totally reflected.

8. The image in a concave lens is
(a) magnified, (b) diminished, (c) virtual, (d) inverted.

9. On which part of the eyes is the image formed?
(a) iris, (b) cornea, (c) retina, (d) pupil.

10. The image produced by a film projector is
(a) magnified, (b) virtual, (c) real, (d) diminished.

Problems

1. State the laws of reflection. What is the minimum length of vertical plane mirror needed to enable a 1.8 m tall man to see his complete reflected image? [0.9 m]

2. A ray meets a plane mirror at 30° to the normal. The mirror is then rotated through 10° on each side of this position. What is the angle between the reflected rays in the two positions? [40°]

3. An object 24 mm in height stands at right angles to the principal axis of a convex lens at a distance of 36 mm. The focal length of the lens is 12 mm. By drawing rays on a scale diagram determine the position and size of the image. Is the image real or virtual? [18 mm from lens; 12 mm long; real]

4. A driving mirror is made from a cylindrical mirror of radius 120 mm and length (measured round the curve) of 120 mm. Assuming the eye of the driver is a long way from the mirror, what is his angle of view? [2 radians]

5. Determine by drawing the position and size of the image formed by a converging mirror of focal length 80 mm when an object 5.0 mm high is placed 200 mm from the mirror. [133 mm; 3.3 mm]

6. An object is located at 250 mm from a mirror of focal length $+200$ mm. Find the position of the image by scale drawing. [1.0 m from mirror]

7. An object is placed at 500 mm from a convex mirror of focal length 80 mm. Determine the position and nature of the image. [69 mm behind mirror; virtual]

8. Determine the position of the image of a car in a convex driving mirror of focal length 1.0 m when the car is at (a) 10 m, (b) 4.0 m. [(a) 0.91 m behind mirror; (b) 0.80 m behind mirror]

9. A shaving mirror 50 mm from a man's face gives a magnification of 2. Find the focal length of the mirror. [100 mm]

10. A dentist uses a concave mirror of 40 mm radius of curvature 10 mm away from a tooth. Where is the image formed and what is the magnification? [20 mm behind mirror; magnification 2]

11. A light ray strikes a 200 mm thick glass block at 60° to the normal. Calculate the lateral displacement of the ray. [$n_g = 1.5$; 102 mm]

12. If a skin-diver sees an object high in the air at 30° to the vertical, what is its real angle to the vertical? $n_w = 1.3$. [41°]

13. What is the wavelength of light in glass of refractive index 1.50 if its wavelength in air is 5.88×10^{-4} mm? [3.92×10^{-4} mm]

14. A light ray strikes one face of an equilateral glass prism at 50° to the normal. At what angle does it emerge, and what is the deviation produced? $n_g = 1.5$. [47°; 37°]

15. A light ray strikes a rectangular glass block at 45° to the normal. What is the thickness of the block if a lateral displacement of 10 mm is produced? $n_g = 1.5$. [30 mm]

16. A camera is fitted with a convex lens, of focal length 20 mm. Find by drawing the position and size of the image of an object 10 mm high placed 30 mm from the lens. [60 mm from lens; 20 mm high]

17. An object 20 mm high is placed at 60 mm from a convex lens of focal length 40 mm. What is the size and position of the image and the magnification? [40 mm; 120 mm from the lens; 2]

18. Repeat problem 17 assuming the lens to be concave. [8 mm; 24 mm from lens; 0.4]

19. A convex lens of focal length 60 mm is used to form an image twice as big as an object. How far is the object from the lens if the image is real and how far if it is virtual? [90 mm; 30 mm]

20. The lens of a camera has an aperture of 4.0 cm. If its focal length is 11.2 cm, what is its f number? [2.8]

Multiple Choice Answers

1(c), 2(ad), 3(a), 4(bc), 5(bd), 6(ad), 7(ab), 8(bc), 9(c), 10(ac).

18.
Basic Electrical Concepts

Your grandfather was probably born by gaslight. He will remember the start of radio broadcasting and the time when there were more steam engines than electric motors. He was probably amazed by simple electrical inventions and curious to know how they worked. Our lives are based on electrical technology which he would never have dreamed of. But we are so familiar with electrical products that we probably do not notice their existence. Familiarity is death to our curiosity and to many people electricity is a mystery they do not understand.

A technician engineer needs to understand and so we must look again and build on the experience of electricity we already have.

Effects of Electric Current

We associate all kinds of events and devices with electric currents. Electric light, electric transport, electric sounds, etc. They are too numerous to mention. However, there are only three basic effects of an electric current and all the other applications follow from them:

(a) magnetic effect
(b) chemical effect
(c) heating effect.

The magnetic effect of current is the basis of most electromechanical devices. Near a current there is a magnetic field and this exerts a force on other currents or magnetic materials. (Fig. 18.1.) The presence of magnetic materials such as iron can make the forces thousands of times greater than the currents acting alone, and yet it is the current which controls the magnet. (Fig. 18.2.) Loudspeakers and electric motors are other applications of electromagnetism to be found on pages 176 and 222.

The materials themselves may retain the magnetism and become permanent magnets which exert their own influence. Permanent magnets are

Wire   Coil   Permanent magnet

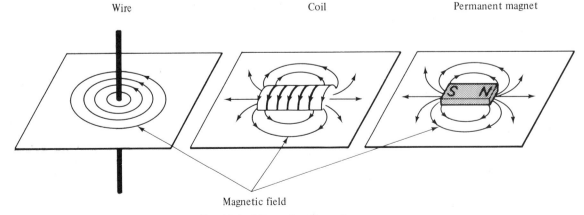

Magnetic field

Fig. 18.1 Magnetic effect of a current

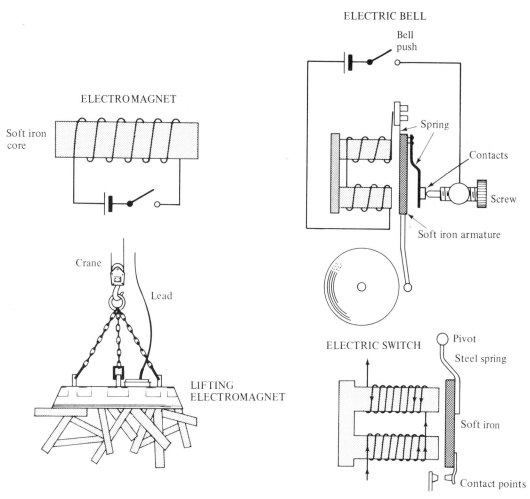

Fig. 18.2　The electromagnet and some applications

The basis of some of the simpler devices. The compass needle responds to the magnetic field of the Earth which is itself a permanent magnet. (Fig. 18.3.)

The Chemical Effect of Current

When a lead–acid battery is charged the acid becomes more concentrated and hydrogen and oxygen are liberated. As the battery discharges the acid gets weaker and lead oxide on the positive plate is changed to lead sulphate. These processes are examples of the chemical effect of a current, i.e., electrolysis. (See page 148.) Electrolysis is used to purify metals such as copper and aluminium and to deposit metals onto surfaces, e.g., silver plating. (Fig. 18.4.)

The Heating Effect of Current

An electric fire is the most obvious example of the heating effect of a current. The element of the fire is just highly resistive wire which glows red hot as the current passes through it. If the wire is very thin it is heated white hot and a greater proportion of light to heat is released as in the tungsten filament lamp. (Fig. 18.5.) Hotter still and more dramatic are the effects of arc welding and fork

197

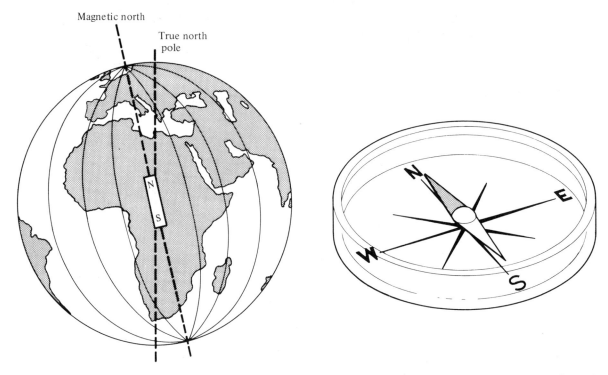

Fig. 18.3 A magnet responds to the horizontal component of the earth's magnetic field which approximates to lines of longitude

lightning when large amounts of electrical energy are concentrated to give temperatures sufficient to melt metals.

There is a problem with electricity which prevents some students from really understanding it. You cannot see it! You cannot see an electron like you can see a drop of water. You cannot see an

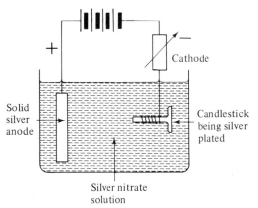

Fig. 18.4 Chemical effect of current. Silver is deposited from the solution onto the cathode

electric current like you can see a jet of water coming from a hose pipe. You cannot feel the voltage exerted by a battery as you can feel the pressure of a pump. If you do find electric current difficult to visualize, try comparing it with the very similar situation of liquid flow. (Fig. 18.6.)

A pump exerts pressure which drives a flow of liquid around the system against the resistance of the pipes.

A battery exerts a voltage which drives a flow of charge around the circuit against the resistance of the wires.

In both cases the rate of flow is the current and unless there is a complete circuit no current can flow. The liquid flows from high to low pressure while the current flows from positive to negative voltage.

To summarize:

In an electric circuit the voltage drives the current through the resistance. It is this basic process which leads to microprocessors, hydroelectric power and the lights of Piccadilly.

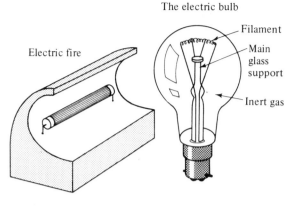

The electric bulb

Electric fire

Filament

Main glass support

Inert gas

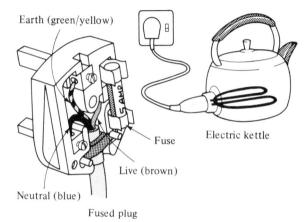

Earth (green/yellow)

Live (brown)

Neutral (blue)

Fuse

Electric kettle

Fused plug

Fig. 18.5   Applications of the heating effect of current

Current, I

An electric current is a stream of charged particles. In a conductor the particles that move are electrons which are so small that they can flow past the atoms without much resistance. (Fig. 18.7.)

Current I is measured in terms of the quantity of charge Q flowing per unit time.

$$I = \frac{Q}{t}$$

The charge on a single electron is inconveniently small as a unit of charge. Quantity of charge is measured in coulombs (symbol c) where 1 coulomb = 6.24×10^8 electronic charges. We can now define the unit of current the ampere (symbol A)

$$1 \text{ ampere} = 1 \text{ coulomb per second}$$

Problem
During 8 seconds 36 coulombs of charge pass a point in a circuit. Calculate the current.

Solution

$$I = Q/t$$

$$I = \frac{36}{8} = 4.5 \text{ A}$$

The current flowing is 4.5 amperes.

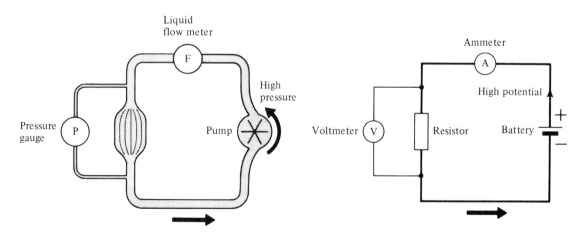

Liquid flow meter

High pressure

Pressure gauge

Pump

Ammeter

High potential

Voltmeter

Resistor

Battery

Fig. 18.6   Comparison with fluid flow helps to fix ideas about circuits

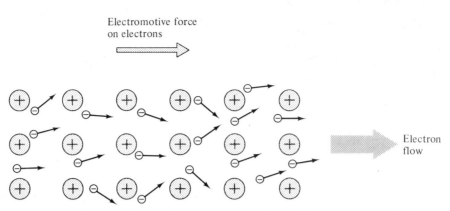

Electromotive force
on electrons

Electron
flow

Fig. 18.7 Electrons move under the influence of an electromotive force. A force in the opposite direction acts on the nucleus but they are fixed

Note that the circuit must be complete to allow a current to flow. The direction of flow of a current is taken to be from positive to negative. Not until after this convention was adopted was it discovered that it is generally the electrons that do the moving. Being negative they actually move from negative to positive! Nonetheless it was decided to stick to the old convention and so you will see conventional current goes from + to − (Fig. 18.8.)

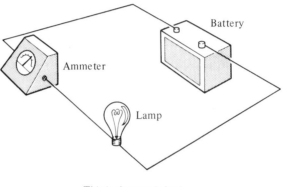

Battery

Ammeter

Lamp

This is the way it looks

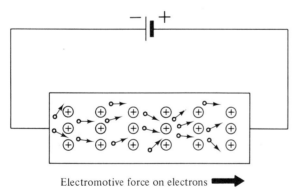

Electromotive force on electrons

Fig. 18.8 Electrons flow around a complete circuit

Current is measured by an ammeter. This is connected in line with the flow. In a series circuit where there is no alternative route for the electrons the current is the same in all parts of the circuit. In this case the ammeter may be connected in the flow at any point in the circuit. (Fig. 18.9.)

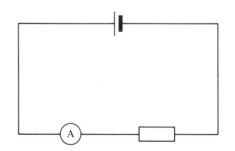

This is the way we represent it on a circuit diagram

Fig. 18.9 An ammeter is connected in series with the circuit

200

Voltage, V

The voltage is the force that drives the current around the circuit. The source of energy, e.g., a battery, exerts the force on the charges by setting up a positive or high potential at one contact, and a negative or low potential at the other. The current flows from the positive to the negative contact or expressed another way, from higher to lower potential.

The potential drops around the circuit from positive to negative just as the liquid pressure drops around the pipes from the high pressure to the low pressure side of the pump of Fig. 18.6. The difference in potential between two points is measured in volts. In fact we use the terms 'voltage' or 'voltage drop' or 'potential difference' (p.d.) to mean the same thing.

Voltage is really a measure of how much energy is available to force each coulomb of charge from one point to another.

$$\text{Voltage} = \text{energy per unit charge}$$

i.e.,

$$\text{Volts} = \frac{\text{joules}}{\text{coulombs}}$$

When we are considering the maximum energy per coulomb that a source of voltage can supply, we refer to it as the electromotive force or e.m.f. for short. The e.m.f. of a source is measured in volts. E.m.f. is only used to describe a source of potential such as a battery or generator whereas the terms voltage or potential difference (p.d.) may refer to any part of a circuit.

Voltmeters are used to measure potential dif-ferences. They are not connected in line with the circuit but parallel to it between the points that are being considered. (Fig. 18.10.)

Problem

What is the potential difference across a filament lamp if it delivers 2000 joules of energy when 50 coulombs of charge pass through it.

Solution

$$V = \frac{\text{joules}}{\text{coulombs}} = \frac{2000}{50} = 40 \text{ V}$$

The potential difference across the lamp is 40 V.

Resistance, R

Ohm's Law We have seen that a voltage applied to a conductor in a circuit produces a current, but how much current? For some conductors we find that **the current I is directly proportional to the voltage V**, i.e.,

$$I \propto V$$

This is Ohm's law and can be expressed in the experimental results shown in the graph of Fig. 18.11.

The ratio of V to I for a particular conductor is called the resistance of the conductor R

$$R = \frac{V}{I}$$

The unit of resistance is the ohm, symbol Ω. The resistance determines how much current flows for

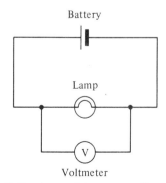

Fig. 18.10   Voltmeter used to measure the potential dif-ference across the contacts of a filament lamp

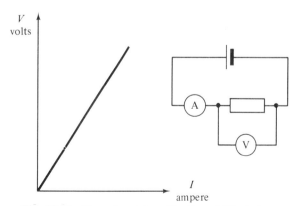

Fig. 18.11   Experimental verification of Ohm's law

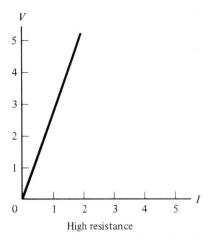

Fig. 18.12 V–I graphs for high and low resistance

a given voltage. Rearranging the equation we have

$$V = IR \quad \text{or} \quad I = \frac{V}{R}$$

A high resistance restricts the current to a low value. When the resistance is small the current is large. (Fig. 18.12.)

Problem

A potential of 6 V is applied to a resistor of 2.5 Ω calculate the current.

Solution

$$I = \frac{V}{R} = \frac{6}{2.5} = 2.4 \text{ A}$$

Current in the resistor = 2.4 A

Problem

Calculate the resistance of a conductor if a voltage of 20 V produces a current of 3.2 A.

Solution

$$R = \frac{V}{I} = \frac{20}{3.2} = 6.35 \text{ Ω}$$

Resistance of conductor = 6.25 Ω

Some resistance is essential in a circuit to limit the current and safeguard the circuit and the source of e.m.f. from damage. Fuses are incorporated in many circuits to keep the current within safe limits. A fuse is a wire which melts and breaks at a designed current value. Alternatively a circuit may incorporate a cut out switch which automatically opens if a certain current is exceeded.

Not all conductors obey Ohm's law. The filament lamp when hot has a higher resistance than when cold. Thus as the current increases and the filament temperature rises, the graph shows a deviation from a straight line. (Fig. 18.13.) There are many non-linear materials which have a $V \sim I$ graph deviating from a straight line, e.g., electrolytes and semiconducting materials. Most metal resistors and carbon act as linear resistors for moderate current values.

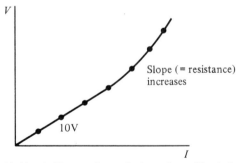

Fig. 18.13 A filament lamp deviates from Ohm's law at high temperatures. Its resistance increases and the slope of the graph increases

Safety First

Remember always that electricity is dangerous. The human body depends on minute electrical

202

currents from the brain to control processes like heartbeat and respiration. An electrical shock may paralyze these processes and stop them for long enough for death to occur. There are no well defined safe limits for electric shocks. Shocks from low voltage car batteries have proved fatal while at the other extreme shocks from high voltage cables do not kill in every case. Beware of people who having survived a number of shocks are scornful of the risks. The statistics show that the 240 V household power supply is lethal under conditions which all too commonly occur.

On the other hand when proper codes of practice are followed and a healthy respect for the dangers is shown then electricity is a valuable and safe element in our lives.

Electrical Circuits and Symbols

Circuit diagrams are always drawn in two dimensions using standard symbols for the components.

Wires or conducting links are represented by lines and their lengths have no significance. All the resistance is assumed to be concentrated in the resistors.

Where lines cross, the conductors do not connect unless that is indicated by a blob at the junction. The direction of current is that of positive or conventional current. Symbols for the more common components are shown in Fig. 18.14. Their meaning will be explained as they arise in the text.

Use of Ammeters and Voltmeters

It is important that meters used to measure current and voltage do not significantly affect the circuit valves. Ammeters must have very low resistances and be connected in series with the circuit. (Fig. 18.15.) On the other hand voltmeters must have a high resistance because they are connected across, i.e., in parallel with parts of the

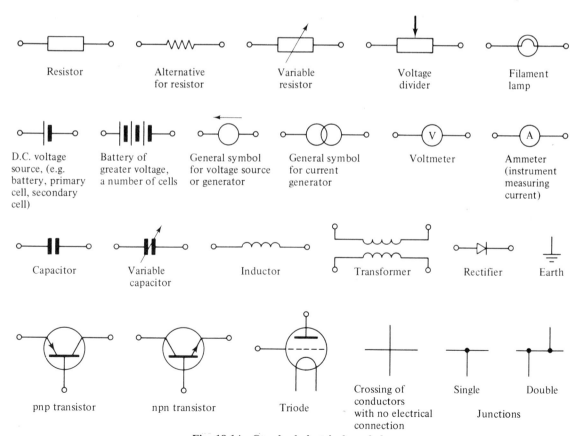

Fig. 18.14 Standard electrical symbols

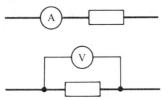

Fig. 18.15   Connect ammeters in series and voltmeters in parallel when measuring circuit values

circuit. They must not offer a low resistance alternative path which would sidetrack significant current from the circuit.

Problem
Calculate the value of R_1 and R_2 in Fig. 18.16 if the current in 4 A and $V_1 = 7$ V and $V_2 = 5$ V.

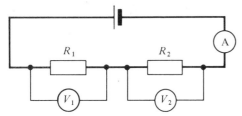

Fig. 18.16   Series resistors

Solution

$$R_1 = \frac{V_1}{I} = \frac{7}{4} = 2.75\ \Omega$$

$$R_2 = \frac{V_2}{I} = \frac{5}{4} = 1.25\ \Omega$$

Series Circuits

When the resistors are connected in series with the e.m.f. source, the same current flows through all of them. (Fig. 18.17.) The total voltage across the

three resistors is equal to the sum of the voltages across the separate resistances R_1, R_2 and R_3. This follows from the fact that voltage is defined in terms of energy and energy is conserved

$$V = V_1 + V_2 + V_3$$

If we divide through by the current I then

$$\frac{V}{I} = \frac{V_1}{I} + \frac{V_2}{I} + \frac{V_3}{I}$$

But from the definition of resistance

$$R = \frac{V}{I}$$

Hence

$$R = R_1 + R_2 + R_3$$

Thus the total resistance of a series circuit is the sum of the separate resistances in the circuit.

Problem
What is the combined resistance of resistors of 12 Ω, 8 Ω and 6 Ω connected in series.

Solution
$$\text{Total resistance} = R = R_1 + R_2 + R_3$$
$$= 12 + 8 + 6 = 26\ \Omega$$

the combined resistance is 26 Ω.

Parallel Circuits

When resistors are connected so as to provide alternative paths for the current we say they are connected in parallel. (Fig. 18.18.)

Again because of energy conservation the same charge taken between the same two points along any of the paths must involve the same work.

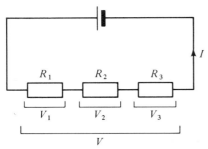

Fig. 18.17   Series circuit. All the current passes through each resistor

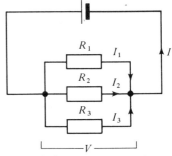

Fig. 18.18   Parallel circuit. The current divides and part goes through different resistors

Therefore the voltage V is common to all three parallel resistors.

The current I divides into components I_1, I_2 and I_3 such that

$$I = I_1 + I_2 + I_3$$

dividing by the common voltage V

$$\frac{I}{V} = \frac{I_1}{V} + \frac{I_2}{V} + \frac{I_3}{V}$$

From the definition of resistance

$$R = \frac{V}{I}$$

for all or any part of a circuit and

$$\frac{1}{R} = \frac{I}{V}$$

Hence

$$\frac{1}{R} = \frac{1}{R_1} + \frac{1}{R_2} + \frac{1}{R_3}$$

Problem

What is the combined resistance of resistors of 3 Ω, 4 Ω and 2 Ω connected in parallel.

Solution

Let R be the value of the combined resistance

$$\frac{1}{R} = \frac{1}{R_1} + \frac{1}{R_2} + \frac{1}{R_3}$$

$$\frac{1}{R} = \frac{1}{3} + \frac{1}{4} + \frac{1}{2}$$

$$\frac{1}{R} = \frac{4+3+6}{12} = \frac{13}{12}$$

$$\therefore R = \frac{12}{13} \, \Omega$$

The combined resistance is 0.92 ohm.

Combined Series and Parallel Circuits

Problems in combined series and parallel circuits test your understanding of Ohm's law and the nature of electrical qualities. It is rather like detective work where you stay cool and methodically use the facts at your disposal to deduce the whole picture.

1. The equation $V = IR$ applies to complete circuits or parts of circuits if the appropriate values of V, I and R are used.
2. Resistances in parallel have the the same voltage but different shares of the current.
3. Resistances in series have the same current and their separate voltage sum to the total voltage.

Problem

Find the current in each part of the circuit of Fig. 18.19.

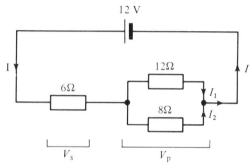

Fig. 18.19   Series and parallel circuit

Solution

First find the total resistance for the parallel resistors.

$$\frac{1}{R} = \frac{1}{R_1} + \frac{1}{R_2} = \frac{1}{12} + \frac{1}{8} = \frac{2+3}{24} = \frac{5}{24}$$

$$R = \frac{24}{5} = 4.8 \, \Omega$$

Adding this to the series resistor

Total resistance $= 4.8 + 6 = 10.8 \, \Omega$

Now we apply $I = V/R$ to the complete circuit to find the total current

$$I = \frac{12}{10.8} = 1.1$$

This passes through the series resistor, hence the voltage across it is

$$V_s = IR = 1.1 \times 6 = 6.6 \text{ V}$$

The voltage across the parallel resistors V_p is difference between the total 12 V and 6.6 V

$$V_p = 12 - 6.6 = 5.4 \text{ V}$$

At last we are in a position to calculate the current in the parallel resistors.

$$I_1 = \frac{V_p}{12} = \frac{5.4}{12} = 0.45 \text{ A}$$

$$I_2 = \frac{V_p}{8} = \frac{5.4}{8} = 0.675 \text{ A}$$

The total current in the circuit is 1.1 A, the current in the 12 ohm resistor is 0.45 A and in the 8 ohm resistor is 0.675 A.

Lamps in Series or Parallel

Lamps are designed to operate on a specific voltage. This power may be supplied from a high voltage source driving current through the lamps in series. Alternatively the power can be supplied from a low voltage driving the lamps in parallel. (Fig. 18.20.)

In each case the lamps are running at their proper voltage so what is the difference? One difference is apparent when one of the bulbs fail. In the series case the circuit is broken and no current flows so all the lamps go out. In the parallel case, only the failed lamp goes out while the others remain alight. The parallel system is therefore much more convenient and is used in the majority of installations. The lights in our homes are wired in parallel and the lamps run on the mains voltage of 240 volts.

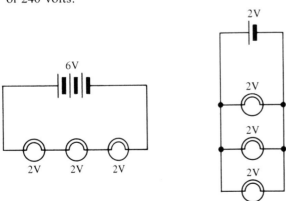

Fig. 18.20 Lamps may be connected in series or parallel provided they receive their rated voltage

When many small light sources are required the series connection has advantages. Small light bulbs are more conveniently made to operate on low voltages. A Christmas tree lights set might include twenty 12 volt bulbs wires in series with a 240 volt supply. Being non-essential lighting we can tolerate a failure. We would use less wire for the series connection and a low voltage interrupter could be used to make the lights flash.

Energy and Power in Electric Circuits

The energy consumed by an electric circuit may be used in doing mechanical work or in storing chemical energy, but it always involves some dissipation of heat. The amount of energy follows from our earlier definition of the volt.

1 coulomb moved through 1 volt needs 1 J
Q coulomb moved through V volts needs VQ J

Remembering that power is the rate of doing work, i.e., power = energy per second.

$$\text{Power} = \frac{VQ}{t} \text{ joules per second}$$

But the current equals the rate of flow of charge, i.e.,

$$I = Q/t$$

Hence

$$\text{Power} = VI \text{ joules per second}$$
$$P = VI \text{ watts}$$

This product is easily remembered when stated as

amperes × volts = watts

It gives the power in any part of the circuit provided the appropriate V and I values are used.

Very often we have values of current and resistance rather than voltage but we can use the Ohm's law expression $V = IR$ to eliminate V from the power equation

$$\text{Power} = VI = IR \times I = I^2 R$$
$$P = I^2 R$$

Problem

An electric fire takes a current of 8 A when connected to a 240 V supply. What power is it consuming?

Solution
$$P = VI = 240 \times 8 = 1920 \text{ W}$$
$$\text{Power} = 1.92 \text{ kW}$$

Problem
Find the power consumed in each of the resistors in Fig. 18.21.

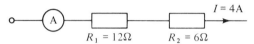

Fig. 18.21   Series resistors

Solution
$$P_1 = I^2 R_1 = 4^2 \times 12 = 192 \text{ W}$$
$$P_2 = I^2 R_2 = 4^2 \times 6 = 96 \text{ W}$$

Problem
Calculate the power dissipated in each of the resistors of Fig. 18.22.

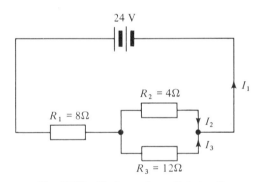

Fig. 18.22   Series and parallel circuit

Solution
We first find the resistance R of the resistors in parallel
$$\frac{1}{R} = \frac{1}{R_2} + \frac{1}{R_3} = \frac{1}{4} + \frac{1}{12} = \frac{3+1}{12} = \frac{4}{12}$$
$$R = \frac{12}{4} = 3 \ \Omega$$

Total resistance in circuit $= 8 + 3 = 11 \ \Omega$.

$$\text{Current } I_1 = \frac{V}{R} = \frac{24}{11} = 2.18 \text{ A}$$

The potential across the parallel resistors
$$V = I_1 R = 2.18 \times 3 = 6.54 \text{ V}$$

The current through R_2,
$$I_2 = \frac{V}{R_2} = \frac{6.54}{4} = 1.635 \text{ A}$$

The current through R_3,
$$I_3 = \frac{V}{R_3} = \frac{6.54}{12} = 0.545 \text{ A}$$

We now have the value of current and resistance for each resistor and using $P = I^2 R$

Power in $R_1 = P_1 = 2.18^2 \times 8 = 38.0 \text{ W}$
Power in $R_2 = P_2 = 1.635^2 \times 4 = 10.7 \text{ W}$
Power in $R_3 = P_3 = 0.545^2 \times 12 = 3.56 \text{ W}$

We could just as easily use $P = VI$

$$P_1 = V_1 I = (24 - 6.54)2.18 = 38.0 \text{ W}$$
$$P_2 = VI_2 = 6.54 \times 1.635 = 10.7 \text{ W}$$
$$P_3 = VI_3 = 6.54 \times 0.545 = 3.56 \text{ W}$$

It is always a relief when it double checks in this way. Note how we use $V = IR$ methodically to find the values in each part of the circuit before calculating P.

The Cost of Electricity

The absolute unit of energy is the joule, but this is inconveniently small for use in the electricity generating industry. A more convenient unit is the energy used when a power of 1 kilowatt operates for one hour. It is called the kilowatt hour (kW h). For example a 3 kW fire burning for 12 hours would consume $3 \times 12 = 36$ kW h. If this were charged at 4p per kilowatt hour

$$\text{total cost} = 36 \times 4 = £1.44$$

Conversion of units from kilowatt hours to joules is not usually necessary, but for the record

$$1 \text{ kW h} = 1000 \times 60 \times 60 \text{ J}$$
$$1 \text{ kW h} = 36 \times 10^5 \text{ J}$$

The kilowatt hour is used only for costings and it should not be used in basic equations for which the joule is the appropriate unit.

Questions

1. Why does a freely suspended bar magnet set itself in an approximately N–S direction?

2. What condition would be necessary for this direction to be exactly N–S at all points of the globe?

3. When a current flows is there a transfer of mass?

4. Why is the conventional direction of current $(+ \rightarrow -)$ different from the flow of the electrons $(- \rightarrow +)$?

5. Where does the comparison of liquid flow with current most help our understanding and where does it break down?

Multiple Choice

1. The coulomb is a unit of what quantity?
   (a) current, (b) charge, (c) resistance, (d) voltage.

2. Which of the following equations does not correctly describe Ohm's law?
   (a) $V = IR$, (b) $I = \dfrac{V}{R}$, (c) $R = VI$, (d) $R = \dfrac{V}{I}$.

3. Which of the following quantities does an ammeter measure directly?
   (a) voltage, (b) current, (c) resistance, (d) charge.

4. Which of the following quantities are measured in volts?
   (a) voltages, (b) potential difference, (c) energy, (d) electromotive force.

5. Which of the following is true for resistors in series?
   (a) $V = V_1 + V_2$, (b) $I = I_1 + I_2$, (c) $R = R_1 + R_2$,
   (d) $R = \dfrac{1}{R_1} + \dfrac{1}{R_2}$.

6. Which of the following quantities are matched with the correct units?
   (a) voltage:joule, (b) current:ampere, (c) resistance:ohm, (d) power:watt.

7. In a parallel arrangement of resistors R_1 and R_2, which of the following are true?
   (a) $\dfrac{1}{R} = \dfrac{1}{R_1} + \dfrac{1}{R_2}$, (b) $I = I_1 + I_2$,
   (c) $V = V_1 + V_2$, (d) $R = R_1 + R_2$.

8. Which of the following are units of electrical energy?
   (a) watt, (b) kilowatt hour, (c) joule, (d) ampere.

9. In a parallel connection of light bulbs to the mains, when one bulb burns out, what is the effect on the other bulbs?
   (a) other bulbs go out, (b) little or no effect, (c) other bulbs go dim, (d) other bulbs go bright.

10. If in question 9 the lamps were in series; what would the effect be?

Problems

1. State Ohm's law. Values of current are measured when different voltages are applied to a resistor. Calculate the resistance in each case and report whether Ohm's law is obeyed.

V	20 V	70 V	105 V
I	6 A	21 A	31.5 A

[3.3 Ω; Ohm's law obeyed]

2. What current is produced when 240 V is applied to a resistance of 60 Ω? [4.0 A]

3. What voltage applied to a resistor of 40 Ω will produce 2.8 A? [112 V]

4. Plot a graph of V against I for a filament lamp from the following data and deduce where deviation from Ohm's law becomes significant. Suggest a reason for this.

Potential in volts
   0   2   4   6   8   10   12
Current in amperes
   0   0.14   0.28   0.42   0.55   0.65   0.74

[Just below 8 volts, heating raises resistance]

5. A resistor carries a current of 0.2 A when a potential difference of 4.0 V is applied across it. What p.d. will produce 0.5 A if Ohm's law is obeyed? [10 V]

6. Calculate the combined resistance of the following combinations
   (a) 5 Ω and 3 Ω in series,
   (b) 12 Ω, 20 Ω and 30 Ω in series,
   (c) 4 Ω and 4 Ω in parallel,
   (d) 4 Ω and 2 Ω in parallel,
   (e) 5 Ω, 10 Ω and 15 Ω in parallel.
   [(a) 8 Ω, (b) 62 Ω, (c) 2 Ω, (d) 1.33 Ω, (e) 2.7 Ω]

7. Three resistors each of 2 Ω are connected
   (a) in series,
   (b) in parallel,
   (c) one in series with a parallel arrangement of the other two. Calculate the combined resistance in each case. [6 Ω; 0.67 Ω; 3 Ω]

8. Calculate the combined resistance of R_1, R_2 and R_3 in Fig. 18.22, where
   (a) $R_1 = 2\,\Omega$, $R_2 = 4\,\Omega$, $R_3 = 6\,\Omega$. [4.4 Ω]
   (b) $R_1 = 5\,\Omega$, $R_2 = 20\,\Omega$, $R_3 = 25\,\Omega$. [16 Ω]
   (c) $R_1 = 1\,\Omega$, $R_2 = 6\,\Omega$, $R_3 = 8\,\Omega$. [4.4 Ω]

9. Calculate the reading of the ammeter in the circuit of Fig. 18.22 for cases (a), (b), and (c), when a potential difference of 1.5 V is supplied by the battery. [(a) 340 mA, (b) 93 mA, (c) 340 mA]

10. Calculate the voltage across R_1 in Fig. 18.22 for cases (a), (b), and (c). [(a) 0.68 V, (b) 0.47 V, (c) 0.34 V]

11. Calculate the voltage across R_2 in Fig. 18.22 and hence determine the current in R_2 for cases (a), (b), and (c).
   (a) [0.82 V, 205 mA]
   (b) [1.03 V, 52 mA]
   (c) [1.16 V, 193 mA]

12. Calculate the power used by a car headlight which takes 3 A from a 12 V supply. [36 W]

13. What current is taken from the 12 V battery of a car when its four parking lights each of 6 W are switched on? [2 A]

14. A bulb rated at 4.5 V and 0.5 A, is to be operated from a 12 V battery. What resistor in series with the bulb would give the required voltage to the bulb? [15 Ω]

15. A 3 kW electric heater operates for 8 hours. What is the cost if the price of electricity is 6p per kW h? [£1.44]

16. How many joules would the heater in problem 15 dissipate during its 8 hours of operation? [86 MJ]

17. Two 60 W 240 V bulbs are connected by mistake in series across the main supply. Calculate the current and power for each bulb assuming Ohm's law is obeyed. [0.125 A; 15 W]

18. Fuses of 2 A, 5 A, 10 A and 13 A are availabe. Which would you use in the following cases using a 240 V supply? (a) 3 kW heater, (b) 400 W electric drill, (c) table lamp of 60 W. [13 A, 2 A, 2 A]

Multiple Choice Answers
1(b), 2(c), 3(b), 4(abd), 5(ac), 6(bcd), 7(ab), 8(bc), 9(b), 10(a).

19.
Electrical Measurements

Measurement of Resistance

The most obvious way of measuring the value of a resistor is to apply a voltage to the resistor and measure the current it produces. (Fig. 19.1.)

Then

$$R = \frac{V}{I}$$

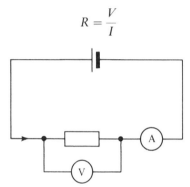

Fig. 19.1 **The value of a resistor obtained by measuring** *VI* **from** $R = V/I$

Resistances may have values ranging from a fraction of an ohm to millions of ohms. When a potential of a few volts is applied to the resistor the current may range from many amps to less than a milliamp. To measure this accurately you need an ammeter with just the right range. For example an ammeter giving a full scale reading of 50 mA would be damaged by a current of 1 A and yet would give almost no deflection for a current of 1 mA.

For many applications an approximate value of resistance is all that we require. In this case we can use a multimeter. (Fig. 19.2.) It applies a standard voltage to a resistor and deflects according to the current produced. There are a number of scales selected by a switch and each scale is calibrated

Fig. 19.2 **A multimeter is a convenient device for measuring resistance**

(*Courtesy of Avo Limited*)

directly in ohms. If you have seen one of these instruments you will know that they have non-linear scales. The scale appears cramped towards one end where small deflections represent large differences in resistance. Hence they give only approximate values for resistance.

In most cases the resistance of a resistor is marked on it in a simple code. (Fig. 19.3.) The

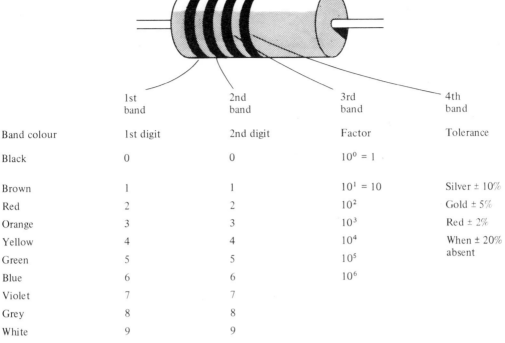

Band colour	1st band 1st digit	2nd band 2nd digit	3rd band Factor	4th band Tolerance
Black	0	0	$10^0 = 1$	
Brown	1	1	$10^1 = 10$	Silver ± 10%
Red	2	2	10^2	Gold ± 5%
Orange	3	3	10^3	Red ± 2%
Yellow	4	4	10^4	When ± 20% absent
Green	5	5	10^5	
Blue	6	6	10^6	
Violet	7	7		
Grey	8	8		
White	9	9		

Fig. 19.3   Colour coding of resistors

code consists of several coloured bands nearer to one end of a resistor. The colour of each band or ring represents a number from 0–9.

The first two bands indicate the first two digits (numbers) of the value. The third ring indicates the power of 10 factor. The fourth ring indicates the tolerance, i.e., the likely limits of accuracy of the quoted value. If there is no fourth band then errors of more than 20 % are possible. (Fig. 19.4.)

Resistivity

How does the resistance of a conductor depend on its dimensions? Exactly as you might expect! The longer it is the more resistance it offers. The greater the cross-sectional area the less resistance it offers, i.e.,

$$R \propto \frac{\text{length}}{\text{cross-sectional area}}$$

$$R \propto l/A$$

If we introduce a constant of proportionality ρ (pronounced rho) we can write

$$R = \rho \frac{l}{A}$$

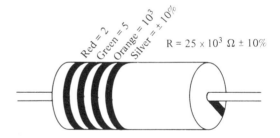

$R = 25 \times 10^3 \ \Omega \pm 10\%$

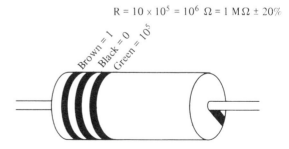

$R = 10 \times 10^5 = 10^6 \ \Omega = 1 \ M\Omega \pm 20\%$

The absence of a 4th band indicates ± 20% tolerance

Fig. 19.4   Colour coding of resistors

The value of ρ is a constant for a particular material and is called the resistivity of that material.

Resistivity values fall into three definite groups

(a) metals, very low ρ
(b) semi conductors, high ρ
(c) insulators, very high ρ

This is illustrated in Fig. 19.5.

Resistors are more accurately made by coiling long lengths of fine metal wire. Less accurate but cheaper and more robust are resistors made from small blocks of carbon. Carbon blocks are also used as self lubricating contacts for electric motors and generators. Semiconducting materials have rather special properties and are used to make transistors in electronic circuits. For most purposes the insulators may be considered not to conduct electric current at all. However, when dampness or when very high voltages are involved conduction may occur. Lightening is an example of insulation breakdown under high voltage. Petrol engines often will not start because of the condensation on the leads to the spark plugs which allows the current to leak away.

Problem

What is the resistance of 200 m of copper wire of diameter 1.0 mm? Resistivity of copper: 1.6×10^{-8} Ωm at 0°C.

Solution

$$R = \rho \frac{l}{A} = \frac{1.6 \times 10^{-8} \times 200}{\pi (0.5 \times 10^{-3})^2}$$

$$= 407.6 \times 10^{-2}$$

The resistance of the wire is 4.08 Ω.

Problem

It is required to construct a resistance of 6.4 ohms from chrome wire of 1.0 mm diameter and resistivity 103×10^{-8} Ωm. What length is required?

Solution

$$R = \rho \frac{l}{A}$$

$$\therefore l = \frac{R \times A}{\rho}$$

$$= \frac{6.4 \times \pi \times (0.5 \times 10^{-3})^2}{103 \times 10^{-8}}$$

Length required = 4.88 m

Table 19.1   Resistivities at 0°C

Material	Resistivity (ohm metres)
Conductors	
Copper	1.6×10^{-8}
Aluminium	2.34×10^{-8}
Iron	9.6×10^{-8}
Gold	2.4×10^{-8}
Silver	1.51×10^{-8}
Zinc	5.5×10^{-8}
Manganin	42×10^{-8}
Nichrome	103×10^{-8}
Carbon	7000×10^{-8}
Insulators	
Ceramics	$>10^{12}$
Glass	$>10^{7}$
Polythene	$>10^{15}$
Paper	$>10^{9}$
Mica	$>10^{11}$
Rubber	$>10^{11}$

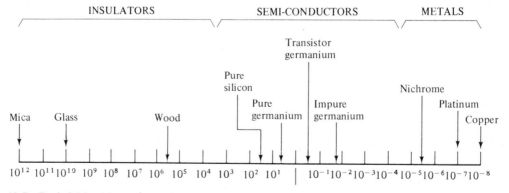

Fig. 19.5   Resistivities (ohm-metre) of conductors, semiconductors and insulators at room temperature (300 K)

Variation of Resistance with Temperature

We have seen that a tungsten filament lamp does not obey the Ohm's law relation $I \propto V$. The reason for this is that for larger currents the filament gets hot and its resistance rises. This increase in resistance with temperature is typical of metals. The increased agitation of the metal atoms causes more disturbance of the electrons flowing past.

Copper increases its resistance three fold for a 500°C change in temperature. (For a more detailed treatment see page 61.)

Unlike metals the resistance of insulators and semiconductors gets less as the temperature rises. (Table 19.2.) At low temperatures there are very few free electrons in insulators and semiconductors, that is why they do not conduct electricity well. At higher temperatures more electrons are separated from the atoms they belong to and so more current is able to flow. This effect is particularly marked in some semiconductors and it is used in devices to control temperature. (Thyristors.)

Table 19.2   Comparison of the Resistivities of Metals, Semiconductors and Insulators at Different Temperatures (figures are very approximate because they vary widely with the method of preparing the sample)

Material	Resistivity		
	0°C	500°C	1000°C
Copper	1.6×10^{-8}	5×10^{-8}	9×10^{-8}
Carbon	3.5×10^{-5}	2.7×10^{-5}	2.1×10^{-5}
Germanium	9×10^{-4}	—	—
Silicon	6×10^{-7}	—	—
Silica	10^{16}	10^7	10^5

Internal Resistance

A battery in a circuit has three distinct functions

1. It applies an e.m.f.
2. It stores energy
3. It is itself a link in the circuit and current passes through it.

So far in the problems and in circuit diagrams we have referred only to one function of a battery, its e.m.f. In many practical situations this simplification is fully justified but not always.

If we want a battery to provide a lot of energy for hour after hour without recharging then it must have a large capacity to store the chemical substances which change as it discharges. Thus the batteries in a calculator giving 12 volts are tiny because the energy requirements are small. Car batteries giving 12 volts are large so that they can operate the lights for long periods of parking before they need recharging. Thus the physical size of batteries is a guide to how much energy they can store.

It is the third function of a battery which we tend to overlook. The battery forms part of the circuit and like every other part it has resistance.

When a current of I amps flows through a resistance of R ohms the voltage drop is IR. When the current flows through the battery itself there is a reduction in voltage of this same product Ir where r is the internal resistance of the battery. (Fig. 19.6.) The actual voltage given by a battery is less than the e.m.f. by an amount Ir

$$V = E - Ir$$

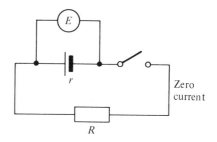

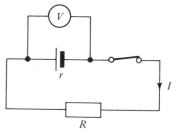

Fig. 19.6   When a current flows the terminal potential difference is less than the e.m.f. $V < E$

When the current and the internal resistance are small the voltage V of a battery is almost the same as the e.m.f. E. When the current is large, then the voltage of the battery is well below its rated e.m.f. Take the case of a car battery with a nominal e.m.f. of 12 V and an internal resistance 0.15 Ω. With the car radio on taking 0.1 A

$$V = E - Ir$$
$$= 12 - 0.1 \times 0.15$$
$$= 12 - 0.015$$
$$= 11.985 \text{ volts}$$

i.e., the difference between the battery voltage and the e.m.f. is insignificant. Now let us switch on the headlights. They take about 10 A.

$$V = E - Ir$$
$$= 12 - 10 \times 0.15$$
$$= 12 - 1.5$$
$$V = 10.5 \text{ volts}$$

The battery voltage is now significantly different from its rated e.m.f.

Finally we switch on the starter motor which takes about 50 A

$$V = E - Ir$$
$$= 12 - 50 \times 0.15$$
$$= 12 - 7.5$$
$$V = 4.5 \text{ volts}$$

The terminal voltage is now only a fraction of its rated e.m.f. You may have noticed that when the starter motor is activated the lights dim and the radio does not work. That is because they are receiving 4.5 volts instead of 12 volts.

Thus if a battery is to supply a large current then its internal resistance must be small. This requires the plates to have large areas. Heavy duty batteries are made up of interleaved plates, the more plates, the greater the area and the less the internal resistance. (Fig. 19.7.) Car batteries have 9, 11, or 13 plates in each cell according to the current required of them.

At the other extreme batteries for calculators and watches take only minute currents of the order of microamps and their internal resistance is not significant. It is usually assumed in circuits that when internal resistance is not mentioned it is insignificant and can be ignored.

Thus to summarize

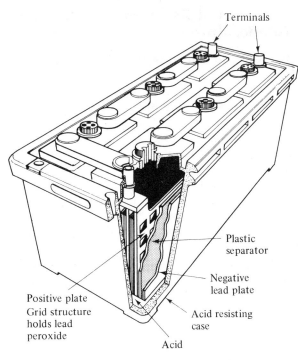

Fig. 19.7 Features of a lead-acid battery
(*Courtesy of Chloride Industrial Batteries Limited*)

E.m.f. is determined by the number and type of cells.

Energy storage capacity $\propto$ the total amount of active chemicals.

Current supplying ability $\propto$ plate area.

When a battery ages the amount of active chemicals on the plates reduces. The effective plate area gets less and the internal resistance rises. This will not be obvious when only a small current is taken and the battery gives its full voltage. When a large current flows however, the terminal voltage of the old battery drops to a low value. An old car battery may well work the lights but when the starter motor takes a heavy current the terminal voltage drops drastically. If this is insufficient to energize the spark to the cylinders, the engine will not start.

Cells in Series and Parallel

The basic source of electromotive force is the cell. When a cell is fully charged it gives an e.m.f. which depends on the type of cell. For example a lead acid cell gives an e.m.f. of 2.2 V. To obtain a higher e.m.f. we connect a number of cells in series, i.e., we make a battery of several cells. The so called 12 volt car battery is a series connection of 6 lead acid cells and gives a total e.m.f. of $6 \times 2.2 = 13.2$ volts.

Note that in the series connections the positive electrode of one cell is connected to the negative electrode of the next. The e.m.f.s are additive. (Fig. 19.8.) Note also that the current goes through each cell in turn and therefore the internal resistance of the battery is the sum of the internal resistances of the separate cells.

Cells connected in parallel have all their positive electrodes connected together at one point and all their negative electrodes connected together at another point. The e.m.f. between these two points is only the same as that of each of the cells taken separately. Why then do we bother to connect cells in parallel?

Several cells connected in parallel have a greater current supplying ability than any of the individual cells used separately. Being connected in parallel their combined internal resistance is less than that of its individual cells so that they can supply a greater current. Also since the combined cells contain a greater amount of active chemical than one cell, the combination can supply a given current for a larger period. Connecting cells in parallel is in effect making a larger cell. Its plate area is larger giving a lower resistance and a larger storage capacity than a single cell.

Potential Divider

We have seen how the e.m.f. can be built up by connecting cells in series. How do we derive a source of e.m.f. which is less than that of an available source, e.g., derive 1 V from a 2 V cell?

We use the fact that there is a difference in potential between two points in a circuit. Take the case of a 12 V battery connected in series with two 50 Ω resistors. (Fig. 19.9.) There is a potential difference across one resistor of 6 V which could be used as a source of potential for some other circuit. We have divided the voltage of the battery in half.

We could divide the battery potential difference in other proportions by changing the ratios of the two resistors.

If one of the resistors is continuously variable then the battery potential can be continuously divided into a range of values up to the limiting voltage of the battery. The potential divider is sometimes called a potentiometer.

Strictly the job of a potentiometer is to measure potential not to divide it. A potentiometer allows us to measure the e.m.f. of a cell directly under conditions in which the cell is delivering no current. Basically the potentiometer is a length of uniform high resistance wire fastened to a metric rule, connected in series with a battery. (Fig. 19.10(a).) A uniform gradient of potential is thus established along the wire. The test cell is connected to the wire at one end and then via a sensitive centre zero ammeter (galvanometer) to a knife edged

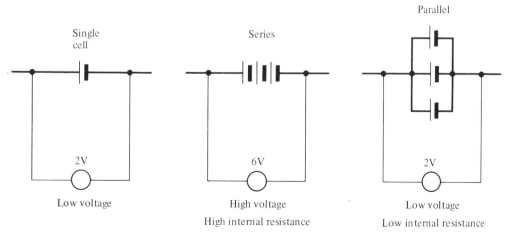

Fig. 19.8   Cells in series and in parallel

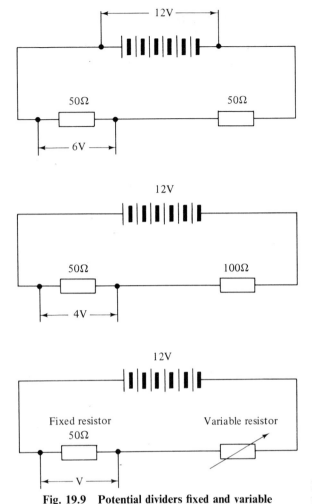

Fig. 19.9 Potential dividers fixed and variable

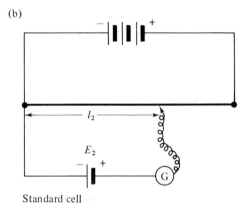

(a)

$$\frac{E_1}{E_2} = \frac{l_1}{l_2}$$

Resistance wire

l_1

E_1

Test cell

(b)

l_2

E_2

Standard cell

Fig. 19.10 A potentiometer measures e.m.f. by comparison with a standard cell

sliding contact. At some contact point along the wire the potential difference is equal to that of the test cell and zero current is recorded in the galvanometer.

e.m.f. ∝ length

Since at balance no current flows through the test cell, its internal resistance is not involved and we are measuring directly its e.m.f.

To convert the balance length l into an e.m.f. the potential gradient (volts/metre) must be known.

This we establish by finding the balance point for a standard cell which gives a fixed e.m.f. to a high degree of accuracy. (Fig. 19.10(b).)

$$\frac{E \text{ test}}{E \text{ standard}} = \frac{l \text{ test}}{l \text{ standard}}$$

Because this determination involves finding the null point for both the standard and test cells it is not very convenient. For most purposes a high resistance voltmeter will measure e.m.f. accurately enough.

Questions

1. Why use coloured bands instead of numbers to indicate the value of resistors?

2. What feature of the sub-atomic structure makes most materials either very good or very bad conductors?

3. Why should hot atoms impede electrons more than cold ones?

4. How is it that an old and new car battery give the same e.m.f. after charging but have very different performances in use?

5. What are the advantages of a potential divider over its alternatives?

Multiple Choice

1. Which band of a code marked resistor shows the tolerance?
(a) 1st, (b) 2nd, (c) 3rd, (d) 4th.

2. Which of the following is directly proportional to the resistance of a uniform wire?
(a) length, (b) radius, (c) cross-sectional area, (d) volume.

3. Which has the highest resistivity?
(a) gold, (b) iron, (c) carbon, (d) polythene.

4. Which materials increase their resistance with temperature?
(a) insulators, (b) plastics, (c) metals, (d) semi-conductors.

5. Which features of a battery allow it to give a high current?
(a) low internal resistance, (b) large plate area, (c) large plate thickness, (d) volume of acid.

6. Which of the following increases the e.m.f. of a combination of cells?
(a) series connections, (b) parallel connection, (c) large plate area, (d) low internal resistance.

7. What factors influence the resistance of a wire?
(a) density, (b) mass, (c) temperature, (d) resistivity.

8. Which of the following describes a lead acid cell?
(a) rechargeable, (b) primary cell, (c) secondary cell (d) not rechargeable.

9. The e.m.f. of four two volt cells connected in parallel is
(a) 2 V, (b) 4 V, (c) 6 V, (d) 8 V.

10. Which combination of similar cells gives the lowest internal resistance?
(a) single cell, (b) two cells in series, (c) two cells in parallel, (d) 3 cells in series.

Problems

1. The voltage across a resistor was measured as 200 volts when the current was 2 milliamps. What is the value of the resistor? [100 kΩ]

2. How does the resistance of a wire vary with (a) length, (b) cross-sectional area? The resistance of a coil of wire of length 200 m, of cross-sectional area 0.5 mm^2 is 20 Ω. Calculate the resistance of 1000 m of the wire of the same material but of cross-sectional area 1.0 mm^2. [50 Ω]

3. Define resistivity. Calculate the resistivity of the wire in problem 2. [$5.0 \times 10^{-8}\ \Omega$ m]

4. Calculate the resistivity of the material of a wire if a 2 metre length of the wire of diameter 0.9 mm has a resistance of 1.2 Ω. [$38 \times 10^{-8}\ \Omega$ m]

5. Manganese wire has a resistivity of $42 \times 10^{-8}\ \Omega$m. What would be the resistance of 6.0 m of manganese wire of diameter 0.46 mm? [15.2 Ω]

6. Calculate the length of Nichrome wire of diameter 0.90 mm and resistivity $103 \times 10^{-8}\ \Omega$ m which would have a resistance of 24 Ω, [14.8 m]

7. A carbon contact has a square section of 8.0 mm side and it is 18 mm long. Calculate the resistance between the ends of the block if the resistivity is 7000 $\times 10^{-8}\ \Omega$ m. [0.020 Ω]

8. Calculate the resistance of 2.0 m of constantan wire of cross-sectional area 1.6×10^{-6} m^2 if the resistivity of constantan is 0.50 $\mu\Omega$ m. [0.63 Ω]

9. A battery of e.m.f. 12.00 V has an internal resistance of 0.18 Ω. What would the battery voltage be when it is delivering a current of 2.0 A? [11.64 V]

10. A torch battery of e.m.f. 4.5 V has an internal resistance of 1.2 Ω. What is the potential difference between its contacts when the bulb takes 0.2 A? [4.26 V]

11. A resistor of 5.0 Ω is connected across the terminals of a battery of e.m.f. 12 V and internal resistance of 0.5 Ω. Calculate the current in the circuit and the voltage across the resistor [2.2 A, 11 V]

12. Plot a graph of terminal voltage against current from the following

Current (A)	1.0	2.0	3.0	4.0	5.0
Terminal voltage (V)	22.0	20.0	18.1	16.3	14.5

Extend the graph to zero current and hence deduce the e.m.f. of the battery. [$\simeq 24$ V]

13. Calculate the internal resistance of the cell in problem 12. [$\simeq 2\ \Omega$]

Multiple Choice Answers

1.(d), **2.**(a), **3.**(d), **4**(c), **5**(ab), **6**(a), **7**(cd), **8**(ac), **9**(a), **10**(c).

20.
Magnetic Fields

Magnets are fascinating objects coming very near to being magical in that they provide our first experience of unseen forces.

Magnetic materials notably iron are attracted by magnets and can themselves acquire magnetism from a magnet without it losing its power. A freely suspended magnet mysteriously sets in the N–S direction. This is the basis of calling the concentration of magnetic influence, the north and south poles of a magnet. These poles react strongly on each other, like poles repelling (N–N and S–S) and unlike poles (N–S) attracting.

Near to a magnet where the magnet exerts an influence we say there is a magnetic field. The strength and shape of a field can be detected by other magnets such as a tiny compass or by magnetic materials such as iron filings. (Fig. 20.1.) Diagrams show the shape of fields by lines of force drawn in the direction of the force exerted on the N pole of a compass.

A breakthrough came in the study of magnets when it was discovered that electric currents also produce magnetic fields. Permanent magnets are fun and have some important applications but the major concern of electrotechnology is with electromagnetism. The simplest field is that due to current in a straight conductor. Its direction is predicted by the 'corkscrew rule'. (Fig. 20.2.) A helical coil (or solenoid) carrying current produces a field rather like that of a bar magnet. (Fig. 20.3.) The presence of an iron core in the coil increases the field by a large factor (e.g. 100 times), but it is still the current which controls the field. Iron with less impurity tends to be soft and because it is easily magnetized and demagnetized it is used in electromagnets. Steel which is iron hardened by

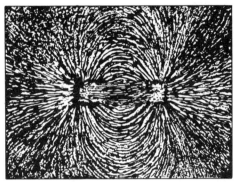

Iron filings show the poles and lines of force of a bar magnet

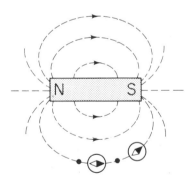

Plotting the field with a compass and representing it by lines of force

Fig. 20.1   Magnetic fields near magnets

carbon impurity, retains its magnetism and is used for permanent magnets.

Magnetic Force on a Current in a Magnetic Field

When a wire carrying a current is placed in a

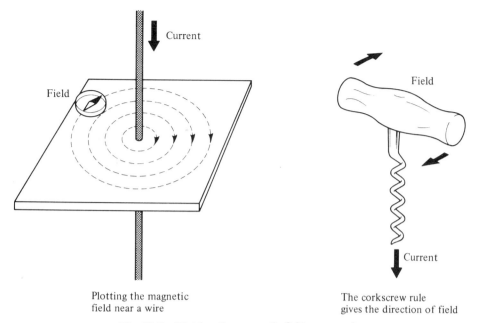

Plotting the magnetic
field near a wire

The corkscrew rule
gives the direction of field

Fig. 20.2   Plotting the magnetic field near a wire

magnetic field it experiences a force. The force acts at right angles to both the field and the current. (Fig. 20.4.) The left hand rule predicts the direction of the force. Careful not to twist your arm or to be misunderstood!

If F is the force on the conductor then

$$F \propto \text{strength of field}$$
$$F \propto \text{length of wire}$$
$$F \propto \text{current in wire}$$

When the conductor is looped within the magnetic field the forces tend to rotate the coil. (Fig. 20.5.) The greater the current which flows the greater the turning force on the coil. Hence we have a means of measuring current. The refinements of a galvanometer (sensitive ammeter) are shown in Fig. 20.6.

The d.c. Motor

The force on the coil of Fig. 20.6 turns it only until the coil is in the vertical position and then it stops. The left hand rule confirms this. In order to make the coil perform a full revolution we need to reverse the current (and therefore the force) at this point. A device for automatically reversing the

current each half revolution is the split ring commutator. (Fig. 20.7.)

This simple coil and commutator would work as an electric motor but it would be weak and it would be jerky. In practice there are several coils wound on an iron core and a commutator with a pair of sections for each coil. The coils receive the current when they are in the strongest part of the field. The current then makes smooth transition from one coil to the next as the brushes transfer to the next sections of the commutator. (Fig. 20.8.)

Electromagnetic Induction

A plane flying above the earth's surface and cutting its magnetic field generates a potential difference of several volts between its wingtips. The plane is an example of a conductor moving relative to a magnetic field and thereby generating an e.m.f. It is only one example of electromagnetic induction, which is the generation of e.m.f., by changing magnetic fields near conductors. We can change the field by several means:

1. Move the conductor
2. Move the field
3. Increase or decrease the strength of field

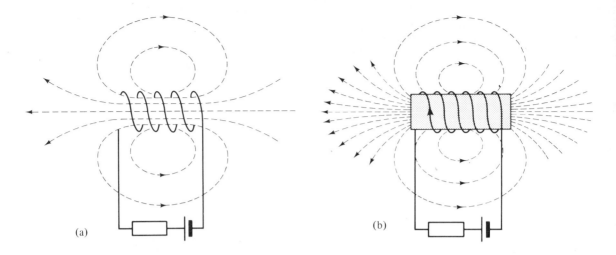

Fig. 20.3 (a) Field due to a solenoid. (b) An electromagnet has an iron core which makes the field many times stronger

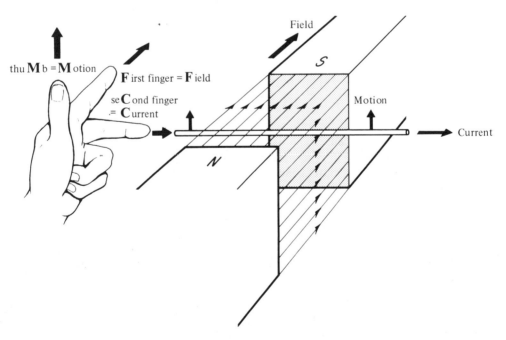

Fig. 20.4 **LEFT HAND RULE. The thumb predicts the direction of the force on the wire. Note the emphasized letters as an aid to memory**

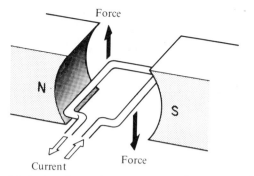

Fig. 20.5 A coil carrying current experiences a turning effect

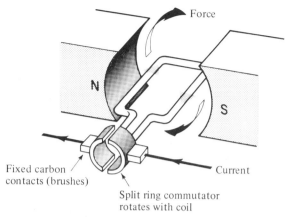

Fig. 20.7 Simple electric motor

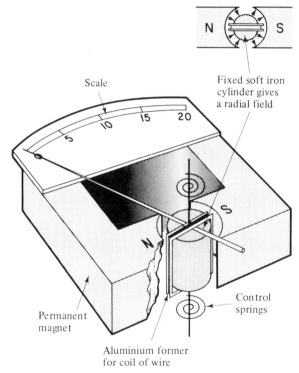

Fig. 20.6 Moving coil galvanometer

These processes are the basis of the generation of e.m.f. by mechanical means. (Fig. 20.9.) The simplest case of a conductor moving across a field is shown in Fig. 20.10. Moving the conductor upwards causes a current to flow in the circuit. Once it has moved across the field the current stops. Moving the conductor downwards causes a pulse of current in the reverse direction. This way of producing short spurts of current is of little value but the electrical generator makes use of it.

The direction of the induced current in a generator may be predicted from the right hand rule. (Fig. 20.10.) (The left hand rule refers to motors, which in Britain we drive on the left!)

The Generator

Forming the conductor into a loop which can be rotated in the magnetic field is the first step towards a generator. (Fig. 20.11.) During one half revolution the current flows towards one of the slip-ring contacts. During the next half revolution the current reverses and flows towards the other contact. The result is alternating current (a.c.) (Fig. 20.11.) Note that the e.m.f. is a maximum when the loop is cutting the central part of the field at right angles. The e.m.f. is zero when the coil is vertical and the wires move parallel to the magnetic field.

We did not set out to get a.c. current, but got it without trying. If we want direct current (d.c.) we can exchange the slip rings for a split-ring commutator so that the brushes change contacts as the current in the coil reverses. The result is a rather irregular direct current e.m.f. (Fig. 20.12.)

A d.c. generator is sometimes called a dynamo. In its practical form the dynamo has several coils each with a corresponding pair of segments on the commutator. The larger the number of coils the less is the irregularity (ripple) in the e.m.f. (Fig. 20.13.) The coils are wound on a soft iron commutator (rotor) to increase both the magnetic field and the e.m.f. generated.

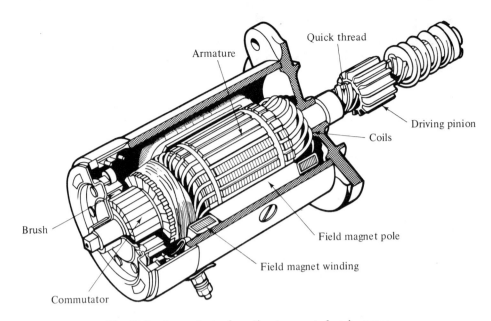

Fig. 20.8   A car starter in a direct current electric motor

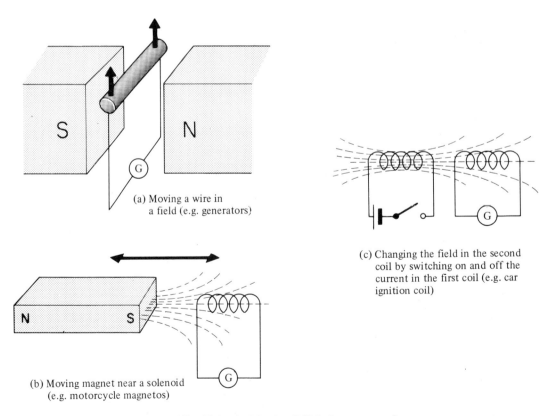

(a) Moving a wire in
a field (e.g. generators)

(b) Moving magnet near a solenoid
(e.g. motorcycle magnetos)

(c) Changing the field in the second
coil by switching on and off the
current in the first coil (e.g. car
ignition coil)

Fig. 20.9   A changing field induces an e.m.f.

222

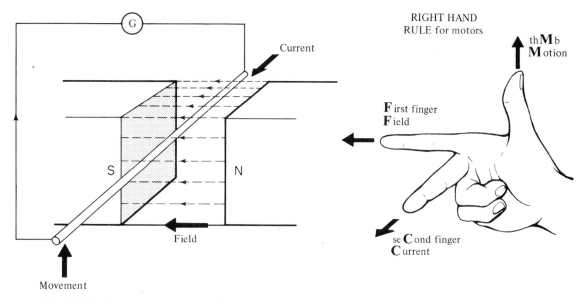

Fig. 20.10   **A moving conductor induces an e.m.f. whose direction follows the RIGHT HAND RULE**

Practical Features of Generators

Only in the very simplest generators and motors is the field provided by permanent magnets. More often electromagnets are used which may be connected in parallel or a series with the coils on the armature (rotor). (Fig. 20.14.)

Shunt wound generators give a steady output but the e.m.f. drops as a heavy current is delivered. By connecting some of the field coils in series (compound winding) the heavier series current to the coils compensates for the falling shunt current and maintains the output e.m.f. Entirely series wound field coils are not now used.

Motors and generators pose a problem because the iron core of the armature is moving in a magnetic field and it is itself a conductor. Hence currents could flow in the armature. These currents called eddy currents would produce over heating and lowered efficiency. Yet without the armature the magnetic fields would be feeble and the efficiency of the machines would be low. How can we preserve the magnetic properties of the armature and eliminate the conduction? The solution is to build the armatures of layers of iron separated by layers of insulation. Eddy currents cannot circulate inside these laminations and the losses are avoided. (Fig. 20.15.)

Back E.m.f.

Experiment shows that a d.c. motor will act as a dynamo if it is forcibly turned. You might expect this since their construction is so similar. But think a little further. If the motor produces an e.m.f. when it is hand turned why should it not produce an e.m.f. when it is self-turned? The fact is that it does. As a motor supplied with a certain voltage increases its speed, so it generates a reverse voltage (back e.m.f.) which tends to oppose the supply. This reduces the current that the motor takes which saves power and also limits its speed. At low speeds when the motor is under load the back e.m.f. is small and the motor accepts more power from the supply. The value of the back e.m.f. is an important element in the design of motors.

Transformers

After generation the electrical power may not be at a voltage suitable for a particular purpose. A transformer changes the voltage of a.c. current up or down as required. Basically the transformer consists of two coils near to each other so that the varying current in one coil, the primary, produces a changing magnetic field in the other, the secondary. (Fig. 20.16.) The varying field induces an

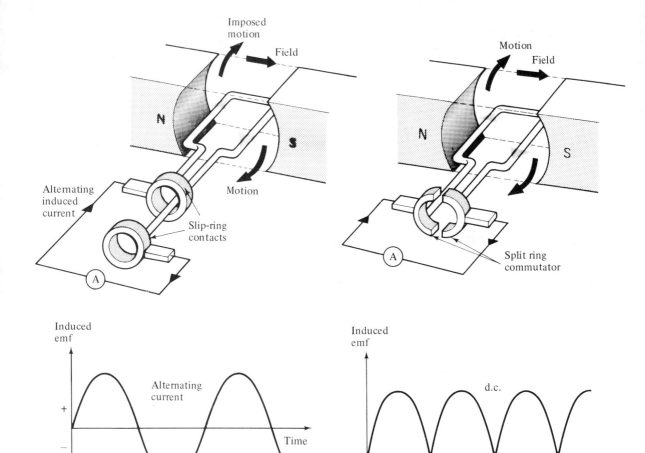

Fig. 20.11 A.c. generator. A loop rotating in a magnetic field generates a current

Fig. 20.12 D.c. generator (dynamo). The split ring commutator switches the contacts each half revolution to give d.c.

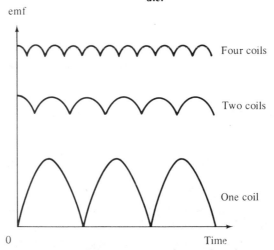

Fig. 20.13 The ripple on induced d.c. currents reduces when the armature has more coils

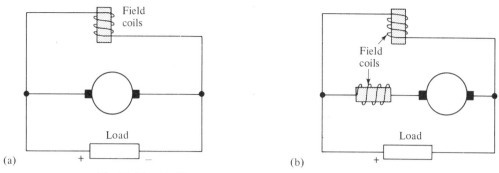

Fig. 20.14   (a) Shunt wound. (b) Compound wound (shunt + series)

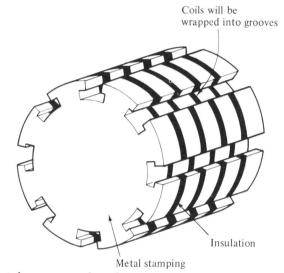

Coils will be
wrapped into grooves

Insulation

Metal stamping

Fig. 20.15   A laminated armature made up from stamped out metal discs separated by insulation

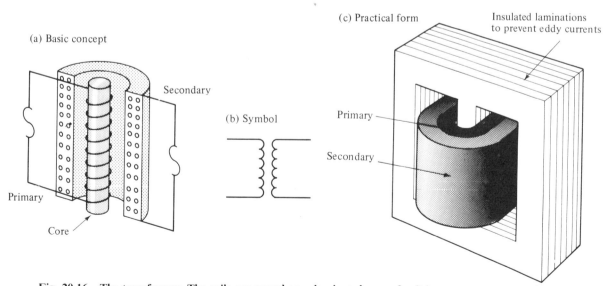

(a) Basic concept

Secondary

(b) Symbol

Primary

Core

(c) Practical form

Insulated laminations
to prevent eddy currents

Primary

Secondary

Fig. 20.16   The transformer. The coils are wound on a laminated core of soft iron to prevent energy losses

alternating e.m.f. in the secondary coil. The frequency of the secondary e.m.f. is the same as that of the primary but the voltage is changed in the ratio of the number of primary to secondary turns.

$$\frac{V_1}{V_2} = \frac{N \text{ primary}}{N \text{ secondary}}$$

For example to increase the voltage tenfold the secondary coil should have ten times as many turns as the primary.

Electrical power is conveyed through the UK national grid at high voltages (a.c.) but transformed to 240 V in local substations for use by consumers.

Changing the voltage of a d.c. supply is more difficult because a transformer cannot be used for d.c. Instead we must use the supply in a d.c. motor which drives a d.c. generator designed to generate the required voltage. This arrangement is called a rotary convertor or dynamometer. It is more costly and requires more maintenance than a transformer.

Rectification

Alternating current has advantages over direct current because a.c. generators are simpler and the voltage of a.c. can be changed readily by means of a transformer. When we require d.c., it can be obtained from a.c. by means of a rectifier. This is a device which transmits current one way only. (Fig. 20.17.) During the positive half of the a.c. cycle current flows, but during the negative half it is blocked. Thermionic valve diodes or semiconducting diodes may be used as rectifiers.

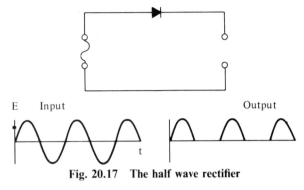

Fig. 20.17   **The half wave rectifier**

Note that in this simple rectifier circuit we lose half the power of the source. If we use four rectifiers we can get a more continuous d.c. supply. Fig. 20.18 shows how a 240 V a.c. supply could be first transformed down to 12 V and then rectified to give 12 V d.c., full wave for a battery charger.

Questions

1. Why is the field of a coil so much greater when it has an iron core?

2. Why should the hard version of iron (steel) be more difficult to magnetize yet keep its magnetism better than the soft pure iron?

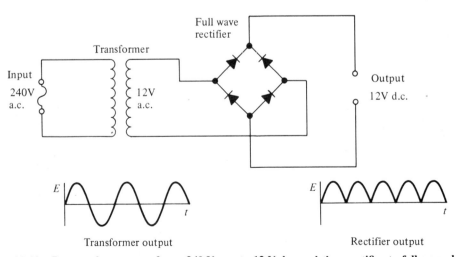

Fig. 20.18   **Battery charger transforms 240 V a.c. to 12 V d.c. and then rectifies to full wave d.c.**

3. The similarity between the field of a bar magnet and that of a solenoid carrying current is too similar for it to be coincidental. What circulating currents might there be in the magnet?

4. Why is the mains electrical supply a.c. and not d.c.?

5. Why do we avoid eddy currents in electrical machines?

Multiple Choice

1. Which of the following statements describe the force between magnetic poles correctly?
(a) like poles repel, (b) all poles attract,
(c) all poles repel, (d) unlike poles attract.

2. The magnetic field of a bar magnet is most like the field due to current in:
(a) a straight wire, (b) a flat coil of many turns,
(c) a solenoid, (d) a single loop of wire.

3. If a fixed current is passed through a solenoid, which of the following core materials would establish the greatest magnetic field?
(a) vacuum, (b) steel, (c) pure iron, (d) aluminium.

4. When the current in question 3 is switched off which core material would display the greatest magnetic field a, b, c or d?

5. Which of the following does **not** influence the force on a conductor in a magnetic field?
(a) field strength, (b) length, (c) cross-sectional area
(d) current strength.

6. A conductor carries current from left to right parallel to the top of this paper while the magnetic field is from bottom to top parallel to the edge of this paper. In what direction will the force on the conductor be?
(a) perpendicular to the paper away from you,
(b) perpendicular to the paper towards you,
(c) from right to left,
(d) from top to bottom parallel to the edge.

7. Which of the following will **not** generate an e.m.f. in a conductor in a magnetic field?
(a) move conductor parallel to the field,
(b) increase the field, (c) decrease the field,
(d) move conductor perpendicular to the field.

8. The purpose of the split ring commutator is to produce
(a) steady current, (b) direct current,
(c) alternating current, (d) maximum current.

9. What is the purpose of a transformer?
(a) to change the voltage, (b) to convert a.c. to d.c.,
(c) to generate electrical power,
(d) to change magnetic fields.

10. Which of the following converts a.c. to d.c.?
(a) a transformer, (b) a rectifier,
(c) a split ring commutator, (d) an armature.

Problems

1. Draw the field due to a bar magnet and due to a current in a solenoid. Use the corkscrew rule to indicate the current direction to the north and south poles of the solenoid.

2. What is the difference in composition of iron and steel. Indicate which of the two materials would be more suitable for the following uses: permanent magnet, lifting magnet, transformer, compass needle, armature.

3. What is the purpose of a transformer? Why is it made up of laminations. The primary coil of a transformer has 40 turns and is supplied with 12 V a.c. If the secondary coil contains 800 turns what will be the secondary voltage? [240 V a.c.]

4. What factors influence the force on a current carrying conductor in a magnetic field? Draw a moving coil meter and indicate how its design might be modified to give an increased deflection for a given current. [increase turns, weaker springs, increase field strength]

5. How does an a.c. generator differ from a d.c. generator? What are eddy currents and how are they reduced in the design of electrical machines?

6. Explain the difference in the input and output of a half wave and full wave rectifier. Give an example of an application where d.c. is essential.

7. Explain by means of a diagram the meaning of shunt and series wound motors. State which is preferred for traction applications.

8. Explain what is meant by back e.m.f. What is its value and effect likely to be on a motor running at (a) low speeds, (b) high speeds?

Multiple Choice Answers

1(ad), **2**(c), **3**(c) **4**(b), **5**(c), **6**(b), **7**(a), **8**(b), **9**(a), **10**(bc).

TEC Standard Units

(Acknowledgement is due to the Technician Education Council for permission to reproduce material from TEC standard units U80/682 and U76/064. The Council reserves the right to revise the content of TEC standard units U80/682 and U76/064 at any time.)

Physical Science I

TEC U80/682
Physical Science I

UNIT CONTENT:

The unit topic areas and the general and specific objectives are set out below, the unit topic areas being prefixed by a capital letter, the general objectives by a non-decimal number, the specific objectives by a decimal number. THE GENERAL OBJECTIVES GIVE THE TEACHING GOALS AND THE SPECIFIC OBJECTIVES THE MEANS BY WHICH THE STUDENT DEMONSTRATES HIS ATTAINMENT OF THEM. Teaching staff should design the learning process to meet the general objectives. The objectives are not intended to be in a particular teaching sequence and do not specify teaching method but, for example practical work could be the most appropriate teaching method for the achievement of the objectives.

ALL THE OBJECTIVES SHOULD BE UNDERSTOOD TO BE PREFIXED BY THE WORDS: THE EXPECTED LEARNING OUTCOME IS THAT THE STUDENT:

A. MATERIAL PROPERTIES AND STATICS

1. Describes in simple terms, the chemical processes involved in burning and rusting as examples of chemical reactions.
   1.1 States that air is a mixture mainly of oxygen and nitrogen.
   1.2 Describes how a substance such as copper gains mass when heated in air and that oxygen is taken from the air by the copper.
   1.3 Describes chemical reactions as interactions between substances which involve a re-arrangement of atoms.
   1.4 States that substances burning in air combine with oxygen, recognising it as an example of a chemical reaction.
   1.5 Describes an oxide as a compound of an element and oxygen.
   1.6 Describes how oxygen and water are involved in rusting, recognising it as a chemical reaction.
   1.7 Gives examples of the damage done by rusting, and a method used to prevent rusting.
2. Describes the elasticity of materials in terms of Hooke's Law.
   2.1 Determines, experimentally, the relationship between force and extension, for different given materials, e.g. rubber, thin wire.
   2.2 States Hooke's Law.
   2.3 Solves simple problems involving Hooke's Law.
3. Solves problems involving coplanar forces in static equilibrium situations.
   3.1 Distinguishes between scalar and vector quantities and gives examples of such quantities.

3.2   States that force is a vector quantity.

3.3   Defines and calculates the moment of force about a point.

3.4   States the "Principle of Moments".

3.5   Solves simple problems using the principle of moments (excluding considerations of reactive forces).

3.6   Defines centre of gravity.

3.7   Determines, experimentally, the centre of gravity of an irregular sheet.

3.8   Describes stable, unstable, and neutral equilibrium.

4. Knows the principles of pressure in fluids.

4.1   Defines pressure.

4.2   Distinguishes between absolute and gauge pressures.

4.3   Calculates pressure given force and area.

4.4   States that the pressure at any level in a fluid is equal in all directions.

4.5   States that the pressure acts in a direction normal to its containing surface.

4.6   States that the pressure due to a column of liquid depends upon the density of the liquid and the height of the column.

4.7   Measures gas pressure using:—
  (a) a U-tube manometer, noting the effect of inclining one limb of the U;
  (b) a pressure gauge.

B.   MOTION AND ENERGY

5.   Solves problems involving distance, time, velocity and acceleration.

5.1   Defines speed.

5.2   Calculates average speed from time and distance data.

5.3   States the difference between speed and velocity.

5.4   Determines, experimentally, distance-time data, plots distance-time graphs, determines the gradients of such graphs and interprets the slopes as speeds.

5.5   Defines acceleration.

5.6   Calculates the gradient of velocity-time graphs and interprets the slope as acceleration.

5.7   Solves problems using the equation: distance = average velocity × time.

5.8   Determines distance from velocity-time graphs.

5.9   States that acceleration of a body is the result of a net force being applied to it.

5.10 States that gravitational force exists and, in the absence of other forces, causes constant acceleration.

5.11 States that frictional forces oppose relative motion between two surfaces.

6. Describes wave motion and solves problems involving wave velocity.

6.1   Lists simple examples of wave motion.

6.2   Explains, using a simple diagram, the meaning of wavelengths and frequency.

6.3   Solves simple problems using $v = f\lambda$.

7. Solves problems associated with energy.

7.1   Defines work in terms of force applied and distance moved.

7.2   Draw graphs, from experimental data, of force against distance moved and hence determines the work done.

7.3   Identifies the forms of energy involved and the conversions occurring in a given system.

7.4   Defines efficiency in terms of energy input and output.

7.5   States that power is the rate of transfer of energy.

7.6   Solves simple problems involving work, energy, efficiency and power.

8. Shows a basic understanding of heat and temperature.

8.1   Defines specific heat capacity.

8.2   Solves problems associated with mass, specific heat capacity and temperature change.

8.3   Describes and experimentally determines the form of the temperature-time graph for a substance changing state, e.g. a liquid cooling, solidifying and then the solid further cooling.

8.4   Differentiates between sensible heat and latent heat.

8.5   States that materials contract or expand with temperature change.

C. ELECTRICITY

9. Solves problems related to current, potential difference and resistance for simple series and parallel resistive circuits.
    9.1 Uses standard symbols for electrical components when drawing circuit diagrams.
    9.2 States that for a current to flow between two points in a circuit a potential difference is required between them.
    9.3 Measures current using an ammeter and potential difference using a voltmeter, and draws a graph of the relationship between potential difference and current, for:—
        (a) a single resistor
        (b) a non-linear component such as a lamp.
    9.4 Defines resistance as the ratio of potential difference (voltage) across a resistor, to the current through it and describes resistance as that property of a conductor that limits current.
    9.5 States Ohm's Law and solves simple problems.
    9.6 Recognises, given a series circuit diagram, that (a) the current is the same in all parts of the current, (b) the sum of the voltages is equal to the total applied voltage.
    9.7 Derives the equation for resistors connected in series, and solves simple problems including the use of Ohm's Law.
    9.8 Recognises, given a parallel circuit diagram, that (a) the sum of the currents in the resistors is equal to the current flowing into the network, (b) the potential difference is the same across the resistors.
    9.9 Derives the equation for resistors connected in parallel and solves simple problems by the use of Ohm's Law.
    9.10 States the relationship between the resistance of a conductor and its length, cross-sectional area, and resistivity, and solves associated problems.
    9.11 States that resistance varies with temperature.
    9.12 Compares the operation of lamps connected in:—
        (a) series
        (b) parallel.
10. Calculate power in electrical circuits.
    10.1 States that the power produced in a circuit is given by
        $P = IV = I^2R = V^2/R$.
    10.2 Calculate the power dissipated in simple circuits.
    10.3 Calculates fuse values given the power rating and voltage of an appliance.
11. Describes chemical effects of electricity.
    11.1 Recognises which materials are good and which bad conductors of electricity.
    11.2 Explains conduction in electrolytes as due to ions.
    11.3 Explains the principle of electrodeposition of metals.
    11.4 Describes a simple cell as two dissimilar conductors in an electrolyte.
    11.5 Explains the difference between primary and secondary cells.
    11.6 Investigates the variation of source potential difference with time for a cell when different loads are used, or compares the performances of different types of cell or investigates the charge/discharge of a secondary cell.
12. Describes electromagnetic effects and simple applications.
    12.1 States that a current-carrying conductor produces a magnetic field and gives an example of an application of this effect, e.g. an electromagnet.
    12.2 Describes the type of magnetic field pattern produced by (a) a bar magnet, (b) a solenoid.
    12.3 States that a current-carrying conductor experiences a force when in a magnetic field.
    12.4 Explains, qualitatively, the basic operation of a moving coil meter.
    12.5 Describes electromagnetic induction with reference to the movement of a magnet in a coil connected to a meter.

Engineering Science I

UNIT CONTENT

The unit topic areas and the general and specific objectives are set out below, the unit topic areas being prefixed by a capital letter, the general objectives by a non-decimal number, the specific objectives by a decimal number. THE GENERAL OBJECTIVES GIVE THE TEACHING GOALS AND THE SPECIFIC OBJECTIVES THE MEANS BY WHICH THE STUDENT DEMONSTRATES HIS ATTAINMENT OF THEM. Teaching staff should design the learning process to meet the general objectives. The objectives are not intended to be in a particular teaching sequence and do not specify teaching method but, for example, practical work could be the most appropriate teaching method for the achievement of the objectives.

ALL THE OBJECTIVES SHOULD BE UNDERSTOOD TO BE PREFIXED BY THE WORDS: THE EXPECTED LEARNING OUTCOME IS THAT THE STUDENT:

A. MATERIALS

1. Describes the structure of matter.
    1.1 Describes the atom as the basic building block of matter.
    1.2 Describes the molecule as an independent group of atoms bonded together.
    1.3 Explains the terms elements and compounds in terms of atomic composition.
    1.4 Distinguishes compounds from mixtures.
    1.5 Gives three examples of each of the following:
        (a) elements
        (b) compounds
        (c) mixtures.
    1.6 Defines the phases of matter, i.e.
        (a) solids
        (b) fluids
        and subdivides fluids into liquids and gases.
    1.7 Gives examples of the conversion of matter from one phase to another.
2. Understands the nature of metallic substances.
    2.1 States five properties characteristic of metals.
    2.2 Lists five typical metallic elements.
    2.3 Explains why metals are rarely used in the pure state.
    2.4 Defines an alloy.
    2.5 Names two alloys composed of metallic elements.
    2.6 Names two alloys which include a non-metallic element.
3. Describes oxidation and reduction.
    3.1 States that air is a mixture mainly of oxygen and nitrogen.
    3.2 Describes how material gains mass when heated in air and that oxygen is taken from the air by the element.
    3.3 Defines oxidation.
    3.4 Uses simple chemical formulae to illustrate oxidation.
    3.5 Defines reduction as the converse process to oxidation.
    3.6 Gives two examples of the obtaining of a metal from an oxide ore.
    3.7 Gives two examples of the presence of impurities in a metal obtained from an oxide ore.
4. Understands the nature of non-metallic materials.
    4.1 Lists typical naturally occurring non-metallic materials.
    4.2 Compares 'natural' with 'man-made' non-metallic materials.
    4.3 Defines a plastics material.
    4.4 Defines a polymer.

4.5   Distinguishes between thermoplastics and thermosetters and gives two examples of each.

4.6   Compares the general properties of plastics and metallic materials.

B. DYNAMICS AND ENERGY

5.   Understands fundamental and derived metric units.
    5.1   States the fundamental units of length, mass and time in SI.
    5.2   States the values of the prefixes: pico, nano, micro, milli, kilo and mega.
    5.3   Defines density and its units.
    5.4   Defines relative density.
    5.5   Solves simple problems relating to mass, volume and density of solids.
    5.6   States that a litre is a cubic decimeter.
    5.7   Measures the density of solids and of liquids.
6.   Solves problems involving distance, time, velocity and acceleration.
    6.1   Defines speed.
    6.2   Calculates average speed from given time and distance data.
    6.3   States the difference between speed and velocity.
    6.4   Plots straight-line distance-time graphs.
    6.5   Calculates the gradient of such graphs and interprets the slope as speed.
    6.6   Defines acceleration.
    6.7   Plots straight-line velocity-time graphs.
    6.8   Calculates the gradient of such graphs and interprets the slope as acceleration.
    6.9   Solves problems using the equation distance = average velocity × time.
    6.10   Calculates distance from the area under velocity-time graphs.
    6.11   Defines the radian.
    6.12   Converts revolutions and/or parts of a revolution to radians and vice-versa.
    6.13   Converts a given angular velocity in rev/min to rad/s and vice-versa.
    6.14   Draws angular velocity-time graphs for motion with constant angular acceleration.
    6.15   Solves simple problems involving distance, time, linear velocity, angle turned, uniform linear and uniform angular acceleration, using appropriate diagrams.
7.   Solves problems involving mass, force, acceleration, area and pressure.
    7.1   Defines force.
    7.2   Defines momentum as the product of mass and velocity.
    7.3   States that the formula $F = kma$ is a mathematical representation of Newton's law that force is proportional to the rate of change of momentum.
    7.4   Uses the formula in 7.3 to define the newton as the unit of force.
    7.5   Solves simple problems involving force, mass and acceleration.
    7.6   States that gravitational force exists and leads to 'free-fall' acceleration.
    7.7   States that the average acceleration due to gravity is approximately 9.81 m/s^2.
    7.8   States that weight is the effect of gravity on a mass, and that the weight of one kilogramme is approximately 9.81 N.
    7.9   Defines intensity of pressure as force per unit area.
    7.10   States that the fundamental derived unit is the newton per square metre and is called the pascal.
    7.11   States that there is a pressure due to the atmosphere.
    7.12   Defines the bar as 100 000 pascals (100 000 N/m^2).
    7.13   States that absolute pressure = gauge pressure + atmospheric pressure.
    7.14   Solves simple problems involving force, area and pressure.
8.   Solves problems involving work, energy and power.
    8.1   Defines work done in terms of force and distance moved.
    8.2   Defines the joule.
    8.3   Defines torque.
    8.4   Defines work done in terms of torque and angle turned.

8.5 Draws graphs of force-distance and torque-angle turned and relates the area under the graph to work done.

8.6 Describes energy as a capacity to do work.

8.7 Names five different forms of energy.

8.8 Gives examples of energy conversion devices.

8.9 States the law of conservation of energy.

8.10 Defines efficiency in terms of energy input and energy output.

8.11 States that power is the rate of doing work or the rate of transfer of energy.

8.12 States that the watt is the unit of power.

8.13 Solves simple problems involving work, energy and power.

9. Solves problems involving mass, specific heat capacity and temperature change.

9.1 Describes fuels as sources of energy.

9.2 States that heat, being a form of energy, the unit for quantity of heat is the joule.

9.3 States that temperature differential decides the direction of transfer of heat energy.

9.4 Distinguishes between temperature and heat energy.

9.5 Defines the fixed points on the Celsius scale of temperature.

9.6 Measures temperature using a mercury-in-glass thermometer.

9.7 Defines a change in enthalpy without change of state (i.e. sensible heat).

9.8 Defines specific heat capacity.

9.9 States that the unit of specific heat capacity is the J/kg°C.

9.10 Determines experimentally the specific heat capacity of water.

9.11 Solves simple problems associated with mass, specific heat capacity and temperature change.

9.12 Defines conduction, convection and radiation.

9.13 Gives one example in each case of the transfer of heat energy by each of the processes given in 9.12.

9.14 Discusses the use of insulation to conserve fuel in a heating installation.

10. Solves problems associated with the turning effect of a force.

10.1 Defines stable, unstable and neutral equilibrium.

10.2 Defines the moment of a force about a point.

10.3 States the principle of moments.

10.4 Defines centroid of a lamina and centre of area.

10.5 Sketches the position of the centroid of a symmetrical lamina such as a rectangle.

10.6 Determines experimentally the position of the centroid of a lamina composed of rectangles.

10.7 Determines experimentally the position of the centroid of a triangular lamina.

10.8 Defines centroid of a mass and refers to the centre of gravity.

10.9 Sketches the position of the centroid of symmetrical solids.

10.10 Determines experimentally the centroid of assymetrical solids.

10.11 Solves simple problems on levers and simply supported beams involving combined concentrated and distributed loads.

C. STATICS

11. Solves problems involving two or three forces meeting at a point.

11.1 Defines a scalar quantity.

11.2 Defines a vector quantity.

11.3 States that force is a vector quantity.

11.4 Uses the notation $F < \theta$ to describe a force.

11.5 Resolves a force into horizontal and vertical components and deduces that $V = F \sin \theta$ and $H = F \cos \theta$, values of θ being restricted to the first quadrant.

11.6 Defines the resultant of two forces meeting at a point.

11.7 States that the components of the resultant are the separate sums of the components of the two forces.

11.8 States the theorem of the parallelogram of forces.

11.9 Solves simple problems involving two forces meeting at a point.

11.10 Defines static equilibrium.

11.11 States that three coplanar forces in equilibrium must meet at a point or act parallel to each other.

11.12 States the theorem of the triangle of forces.

11.13 Solves simple problems involving the triangle of forces with the aid of Bow's notation.

D. ELECTRICITY

12. Discusses the nature of electricity.

12.1 Describes the structure of the atom and defines electron, proton and neutron.

12.2 Describes the 'shells' of electrons in the atom and the detachment of a 'loosely held' electron by some influence.

12.3 States that the free movement of electrons gives a current of electricity.

12.4 States that, for undirectional continuous current, free electrons must be available in all parts of a circuit.

12.5 Defines the terms 'conductor' and 'insulator' and gives three examples of each.

12.6 States that all conductors offer some resistance to the flow of an electric current.

12.7 Discusses the use of primary and secondary cells, and generators, as sources of electricity.

12.8 Selects preferred symbols appropriate to this unit, for the representation of sources, cells, resistors, and switches in an electric circuit.

13. Identifies the three main effects of an electric current.

13.1 States examples of an electric current being used for its magnetic effect.

13.2 States examples of an electric current being used for its chemical effect.

13.3 States examples of an electric current being used for its heating effect.

13.4 Identifies the effect being made use of in given specific cases, e.g. electromagnet, electroplating, electric fire, fuse.

14. Solves problems associated with simple electrical circuits.

14.1 States that the SI unit for current is the ampere.

14.2 States that one ampere is one coulomb per second, that a coulomb is the unit for quantity of electricity and that a coulomb is composed of a specific and very large number of electrons.

14.3 Explains how a current flows due to the existence of a potential difference between two points in an electrical conductor.

14.4 Defines potential difference in terms of energy per coulomb and that the unit is the volt.

14.5 States that electromotive force is the energy per coulomb and that the unit is the volt.

14.6 Shows that since volts = joules ÷ coulombs then joules = volts × coulombs and watts = volts × amperes.

14.7 States Ohm's Law in terms of the proportionality of current to potential difference.

14.8 States that this constant of proportionality for a particular resistor is called the resistance and that the unit of resistance is the ohm.

14.9 States that the reciprocal of resistance is called conductance and that its unit is the siemen.

14.10 Solves simple problems using Ohm's law.

14.11 Describes the difference between series and parallel connections of resistors.

14.12 States that the current is the same in all parts of a series circuit.

14.13 States that the sum of the voltages in an external series circuit is equal to the total applied voltage.

14.14 Shows that for resistors connected in series the equivalent resistance is given by $R = R_1 + R_2 + R_3 + \cdots$

14.15 Solves simple problems involving up to three resistors connected in series, including the use of Ohm's law.

14.16 States that the sum of the currents in resistors connected in parallel is equal to the current flowing into the parallel network.

14.17 States that the potential difference (voltage) is the same across resistors in parallel.

14.18 Shows that for resistors connected in parallel the equivalent resistance is given by:

$$\frac{1}{R} = \frac{1}{R_1} + \frac{1}{R_2} + \frac{1}{R_3} + \cdots$$

and that the equivalent conductance is given by $G = G_1 + G_2 + G_3 + \cdots$.

14.19 Solves simple problems involving up to three resistors connected in parallel by use of Ohm's Law.

14.20 States that the power dissipated in a given conductor is the product of the potential difference and the current.

14.21 Uses Ohm's Law to show that $P = I^2 R$.

14.22 Calculates the power dissipated in simple circuits.

Index